5th International Conference on Stem Cell Engineering 2016

Toronto, Canada
23 - 26 October 2016

ISBN: 978-1-5108-4168-0

Printed from e-media with permission by:

Curran Associates, Inc.
57 Morehouse Lane
Red Hook, NY 12571

proceedings.com

Some format issues inherent in the e-media version may also appear in this print version.

Printed by Curran Associates, Inc. (2017)

For permission requests, please contact AIChE
at the address below.

AIChE
120 Wall Street, FL 23
New York, NY 10005-4020

Phone: (800) 242-4363
Fax: (203) 775-5177

www.aiche.org

Additional copies of this publication are available from:

Curran Associates, Inc.
57 Morehouse Lane
Red Hook, NY 12571 USA
Phone: 845-758-0400
Fax: 845-758-2633
Email: curran@proceedings.com
Web: www.proceedings.com

TABLE OF CONTENTS

DIFFERENTIATION & MORPHOGENESIS I

COMPUTATIONAL STEM CELL BIOLOGY

MICROTECHNOLOGIES & HIGH THROUGHPUT SCREENING/DISEASE MODEL

IN SITU/ENDOGENOUS REGENERATION

IN SITU/ENDOGENOUS REGENERATION

IMMUNE ENGINEERING

CLINICAL APPLICATIONS AND CHALLENGES (INCLUDING SCALE UP AND IMAGING)

*** Formatting Issue - Best Available Paper/Graphic ***

Engineering the Morphogenesis of Pluripotent Stem Cells

5th International Conference on Stem Cell Engineering

- Conference: International Conference on Stem Cell Engineering
- Year: 2016
- Proceeding: International Conference on Stem Cell Engineering 2016
- Group: General Submissions
- Session:

  Differentiation & Morphogenesis I

- Time:

  Monday, October 24, 2016 - 9:00am-9:30am

Authors:
McDevitt, T., Georgia Institute of Technology

Pluripotent stem cells (PSCs) exhibit a unique ability to recapitulate dynamic morphogenic processes that mimic analogous developmental biology events to yield complex tissue structure and function that results from coordinated multicellular interactions. The recent success in generating a diverse set of â□□organoidsâ□□ from PSCs demonstrates their tremendous inherent potential for complex tissue formation. However, the inability to robustly and predictably generate organoids currently limits their use for drug discovery and development, modeling of human development and disease, and regenerative medicine therapies. Thus, we have sought to develop methods to reproducibly and robustly direct the morphogenesis of pluripotent stem cell aggregates to directed microtissues by controlling physical and biochemical elements of the 3D microenvironment. We have systematically examined the effects of numerous factors, such as aggregate formation methods, hydrodynamic forces and presence of various types of biomaterials, on the differentiation and morphogenesis of PSCs. We are also now combining these approaches with novel CRISPR-based cell engineering methods in PSCs to spatially and temporally control the initiation of morphogenic processes and complement our experimental studies with computational modeling approaches. We anticipate that these studies will lead to a greater mechanistic understanding of stem cell biology and tissue development as well as scalable and translatable technologies for manufacturing of PSC-derived products for regenerative medicine.

Dynamic Soft Materials for Deconstructing 'Matrix Structure—Cell Function' Relationships

5th International Conference on Stem Cell Engineering

- Conference: International Conference on Stem Cell Engineering
- Year: 2016
- Proceeding: International Conference on Stem Cell Engineering 2016
- Group: General Submissions
- Session:

  Differentiation & Morphogenesis I

- Time:

  Monday, October 24, 2016 - 9:30am-9:45am

Authors:
Kilian, K., University of Illinois at Urbana-Champaign
Abdeen, A., University of Illinois at Urbana-Champaign

Dynamic soft materials for deconstructing matrix structurecell function relationships

Amr A. Abdeen, Kristopher A. Kilian

Department of Materials Science and Engineering, Department of Bioengineering, Micro and Nanotechnology Laboratory, Institute for Genomic Biology

The extracellular matrix surrounding cells in tissue is a dynamic composite material, where the presentation of biophysical and biochemical information directs functional bioactivities. Here I will discuss how engineered model systems can be leveraged to explore how cells receive and integrate signals to adopt specific bioactivities. We have discovered how engineering actomyosin contractility in single mesenchymal stem cells can normalize the population and activate a medicinal state. Variations in subcellular curvature will guide focal adhesion, cytoskeletal organization, and the regulation of distinct epigenetic marks to orchestrate a medicinal secretome. Within two days, activated cells show elevated expression of pericyte markers, and will recapitulate functional pericyte activity through enhanced association with endothelial cell tubules in co-culture. Enhanced angiogenesis is demonstrated in vivo using micropatterned hydrogels applied to a chick chorioallantoic membrane assay. I will also demonstrate a dynamic magnetically-tunable hydrogel that can reversibly stiffen in response to permanent magnets to modulate stem cell differentiation and pro-angiogenic activity. MSCs are believed to undergo a temporary switch in vivo to an activated state in response to injury; thus we propose engineering actomyosin contractility after isolation can similarly activate MSCs, which may serve as a general approach to prime a medicinal phenotype for improved efficacy of cell-based therapies.

Generating Epicardial Cells from Human Pluripotent Stem Cells in a Defined, Growth-Factor Free Manner

5th International Conference on Stem Cell Engineering

- Conference: International Conference on Stem Cell Engineering
- Year: 2016
- Proceeding: International Conference on Stem Cell Engineering 2016
- Group: General Submissions
- Session:

  Differentiation & Morphogenesis I

- Time:

  Monday, October 24, 2016 - 9:45am-10:00am

Authors:
Bao, X., University of Wisconsin Madison
Lian, X., University of Wisconsin-Madison

The epicardium contributes both multi-lineage descendants and paracrine factors to the heart during cardiogenesis and cardiac repair, underscoring its potential for cardiac regenerative medicine. Despite significant advances in animal models, little is known about the cellular and molecular mechanisms that regulate human epicardial development and regeneration. To assess the roles of developmental pathways in epicardial formation, self-renewal, and differentiation we developed a robust method for differentiating human pluripotent stem cells to epicardial cells via stage-specific modulation of cardiac developmental pathway under chemically-defined, animal component-free conditions. The hPSC-derived epicardial cells were expandable and exhibited properties of primary epicardial cells both in vitro and in vivo. These findings improve our understanding of differentiation and self-renewal mechanisms of the epicardium and have implications for stimulating epicardial regeneration via cellular or small-molecule therapies.

Biomaterials for Cellular Reprogramming and Origami-Based Tissue Engineering

5th International Conference on Stem Cell Engineering

- Conference: International Conference on Stem Cell Engineering
- Year: 2016
- Proceeding: International Conference on Stem Cell Engineering 2016
- Group: General Submissions
- Session:

  Differentiation & Morphogenesis I

- Time:

  Monday, October 24, 2016 - 10:00am-10:15am

Authors:
Hwang, N. S., Seoul National University

Lineage conversion from one somatic cell type to another is an attractive approach for deriving specific therapeutic cell generation.Â We report the development of a direct conversion methodology in which cells are transdifferentiated through a plastic intermediate state induced by exposure to non-integrative minicircle DNA (MCDNA)-based reprogramming factors, followed by differentiation into myoblasts. In order to increase the MCDNA delivery efficiency, reprogramming factors were delivered into the chondrocytes via electroporation followed by poly (Î²-amino esters) (PBAE) transfection. We used this approach to convert human chondrocytes to myoblast, and with treatment of SB-431542, an inhibitor of the activin receptor-like kinase receptors, to enhance myogenic commitment. Differentiated cells exhibited expression of myogenic markers such as MyoD and Myog. This methodology for direct lineage conversion from chondrocytes to myoblast represents a novel non-viral Method to convert hard-to-transfect cells to other lineage. Furthermore, we present a novel method for assembling biofunctionalized paper into a 3-dimensionally (3D) structured complex scaffold system for reliable tissue regeneration using an origami-based approach. The surface of a paper was conformally modified with a poly(styrene-co-maleic anhydride) (PSMa) layer via initiated chemical vapor deposition (iCVD) followed by the immobilization of poly-Llysine (PLL) and deposition of Ca^{2+}. This procedure ensures the formation of alginate hydrogel on the paper due to Ca^{2+} diffusion. Finally, we show that the paper-based scaffold can be folded freely into 3D tissue-like structures by simple origami-based method.

*** Formatting Issue - Best Available Paper/Graphic ***

Using Patient-Derived Stem Cells to Study HIF2 Regulation in Pulmonary Hypertension

5th International Conference on Stem Cell Engineering

- Conference: International Conference on Stem Cell Engineering
- Year: 2016
- Proceeding: International Conference on Stem Cell Engineering 2016
- Group: General Submissions
- Session:

  Differentiation & Morphogenesis I

- Time:

  Monday, October 24, 2016 - 10:15am-10:30am

Authors:
Chan, X. Y., Johns Hopkins University
Black, R., Johns Hopkins University
Fang, L., Johns Hopkins University
Cheng, L., Johns Hopkins University
Lee, F.

Erythrocytosis is a condition in which an abnormally high number of red blood cells (RBCs) are present in the blood. In a healthy person, the production of RBCs are tightly regulated by erythropoietin (EPO), which is controlled by a member of the hypoxia inducible factor (HIF) family, HIF2A. Under normoxic conditions, HIF2A is hydroxylated by prolyl hydroxylase domain protein (PHD), targeting it for degradation. Under hypoxic conditions, this modification of HIF2A is inhibited to stabilize and promotes downstream expression of a variety of target genes, including EPO. As a result of a gain-of-function mutation in the oxygen-sensitive transcription factor HIF2A, a specific group of patients demonstrated erythrocytosis and developed pulmonary hypertension (PH). One main characteristic of PH is endothelial dysfunction, leading to structural changes of pulmonary vasculature. Mouse model with gain-of-function HIF2A mutation demonstrated erythrocytosis and developed PH similar to human patients. However, to date, it is unknown how the HIF2A+ mutation leads to PH, clear of the effects of erythrocytosis. Using human induced pluripotent stem cell (hiPSC) line derived from an erythrocytosis patient and an optimized endothelial cell differentiation protocol we created a human *in vitro* system to study HIF2A+ vascular cells. We found that HIF2A+ hiPSC-derived ECs (VEcad-positive cells) can mature to functional endothelial cellsâ□□expressing mature markers: von Willebrand factor and endothelial nitric oxide synthase are capable of lectin binding and acetylated low-density lipoprotein uptake. Real-time PCR results show that HIF2A mRNA expression is specifically up-regulated in hiPSC-derived ECs. Endothelin-1 (EDN1) expression is significantly higher in HIF2A+ hiPSC-derived ECs compared to the ECs derived from health donor hiPSCs, correlating with the HIF2A gain-of-function mutation. In addition, the expression of ECM proteins including laminin and fibronectin is higher in HIF2A+-ECs compared to the control. We speculated that these differences in gene expression are caused by overactive HIF2A expression, which causes endothelial dysfunction in PH.

*** Formatting Issue - Best Available Paper/Graphic ***

Engineering CNS Tissue Morphogenesis Ex Vivo

5th International Conference on Stem Cell Engineering

- Conference: International Conference on Stem Cell Engineering
- Year: 2016
- Proceeding: International Conference on Stem Cell Engineering 2016
- Group: General Submissions
- Session:

  Differentiation & Morphogenesis I

- Time:

  Monday, October 24, 2016 - 10:30am-10:45am

Authors:
Knight, G., University of Wisconsin-Madison
Marti-Figueroa, C., Wisconsin Institute for Discovery
McNulty, J., University of Wisconsin-Madison
Tokar, J., University of Wisconsin-Madison
Beebe, D. J., University of Wisconsin
Turng, L. S., University of Wisconsin-Madison

The human central nervous system (CNS) is comprised of diverse interconnected tissues, each containing unique cytoarchitectures and cell phenotypes critical to their physiology and associated disease pathologies. Engineering organotypic models of these tissues from human pluripotent stem cells (hSPCs) poses unique challenges because the morphogenetic events that pattern CNS development in vivo are dynamic, combinatorial, and occur at micro- to-millimeter scales. Organoid models to date rely on uncontrolled, spontaneous morphogenesis of EBs or neurospheres that provide limited to no control over the resulting tissue cytoarchitecture. Here, we present novel culture platforms that enable the requisite spatial and temporal control of microscale morphogen gradients and tissue morphology necessary to orchestrate morphogenesis of 2D transverse slice and 3D tube models of the developing spinal cord. These platforms integrate our latest advancements in chemically defined neural differentiation protocols1,2 with micropatterned culture substrates3,4 and sacrificial template-molded hydrogels.

*** **Formatting Issue - Best Available Paper/Graphic** ***

Furthermore, we will describe our advances in implementing such platforms to create a high throughput screening platform for developmental neurotoxicity studies and neurodegenerative disease models.

References:

1. Lippmann, E. S. *et al. Stem Cells* **32,** 1032â□□1042 (2014).
2. Lippmann, E. S. *et al. Stem Cell Reports* **4,** 632â□□644 (2015).
3. Knight, G. T. *et al. Chem. Commun. (Camb.)* **51,** 5238â□□5241 (2015).
4. McNulty, J. D. *et al. Lab Chip* **14,** 1923â□□1930 (2014).

Comprehensive Reconstruction and Manipulation of the Pluripotent Stem Cell Metabolic Network

5th International Conference on Stem Cell Engineering

- Conference: International Conference on Stem Cell Engineering
- Year: 2016
- Proceeding: International Conference on Stem Cell Engineering 2016
- Group: General Submissions
- Session:

  Computational Stem Cell Biology

- Time:

  Monday, October 24, 2016 - 11:15am-11:30am

Authors:
Chandrasekaran, S., Wyss Institute for Biologically Inspired Engineering, Harvard University
Zhang, J.
Daley, G. Q., Children's Hospital Boston
Collins, J. J., Massachusetts Institute of Technology

Metabolism is an emerging stem cell hallmark tied to cell fate, pluripotency and self-renewal. Yet systems-level understanding of stem cell metabolism has been limited by the lack of genome-scale network models. Here we develop a systems approach to integrate time-course extracellular and intracellular metabolomics data with a computational model of metabolism to reconstruct the metabolic state of naïve- and primed- murine embryonic stem cells that have different development potential, and discovered three key insights. First, alterations in one-carbon metabolism involving phosphoglycerate dehydrogenase, folate and nucleotide metabolism are key pathways that differ between the two states, resulting in differential sensitivity to folate inhibitors. Second, the model predicted that the master regulator Lin28 regulates this one-carbon metabolic pathway associated with pluripotency. Third, we identified and validated metabolic reactions such as Adenosylhomocysteinase, that can differentially impact histone methylation in naïve and primed cells. Our network-based approach provides a framework for characterizing stem cell metabolism for rational cellular engineering.

Patterning mesoderm and blood development from human pluripotent stem cells

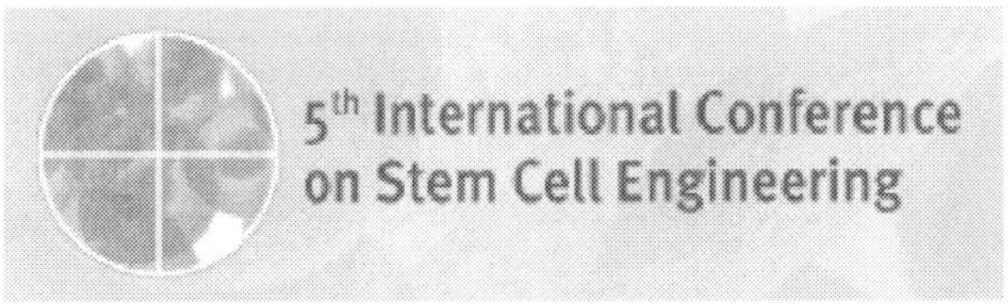

- Conference: International Conference on Stem Cell Engineering
- Year: 2016
- Proceeding: International Conference on Stem Cell Engineering 2016
- Group: General Submissions
- Session:

  Computational Stem Cell Biology

- Time:

  Monday, October 24, 2016 - 11:30am-11:45am

Authors:
Zandstra, P., University of Toronto

Functional tissue emerges from complex spatial-temporal interactions between heterogeneous cell populations. In this presentation I will review our efforts to establish an integrated understanding of the effects of cell type heterogeneity, spatial organization and multi-scale regulatory network engagement on defined and measurable stem cell fate transitions. Specific examples from pluripotent stem cell fate control will be highlighted and new data on modeling human pluripotent stem cell development into definitive blood cells by mimicking the hemogenic niche will be presented.

*** Formatting Issue - Best Available Paper/Graphic ***

Tracking Pluripotent Stem Cell Differentiation with FLOW-MAP, a Graph-Based, Force-Directed Layout Algorithm for Single-Cell, Time Course Datasets

5th International Conference on Stem Cell Engineering

- Conference: International Conference on Stem Cell Engineering
- Year: 2016
- Proceeding: International Conference on Stem Cell Engineering 2016
- Group: General Submissions
- Session:

  Computational Stem Cell Biology

- Time:

  Monday, October 24, 2016 - 11:45am-12:00pm

Authors:
Zunder, E. R., University of Virginia
Ko, M. E.
Rustagi, R. S.
Fragiadakis, G. K.
Nolan, G. P.
Fread, K. I.

Identifying the molecular mechanisms that control the progression of cell differentiation and the branch points of cell fate is an important goal for stem cell biology. Molecular analysis at the single-cell level can provide critical insight toward this goal, because stem cell differentiation occurs in complex and heterogeneous environments where the cell type of interest may be rare and/or transitory. High-content single-cell analysis methods such as flow cytometry, mass cytometry, and single-cell RNA-seq have proven useful for identifying and molecularly characterizing rare cell populations, but using this information to reconstruct cell lineage hierarchies remains challenging. The FLOW-MAP algorithm was developed to serve this need by tracking cell populations as they change over time within single-cell datasets, and combining these trajectories into a global network graph that can be used to infer lineage relationships between cell types. First, the algorithm identifies unique cell populations by density-dependent downsampling and hierarchical clustering. Second, the algorithm connects these cell populations across time points into a sparse graph structure. Third, a force-directed layout function is applied to the graph in order to aid visualization and analysis of the cell lineage trajectories. To investigate the molecular mechanisms that drive cell fate decisions during germ layer formation, this FLOW-MAP approach was applied to an *in vitro* differentiation time course experiment. At Day 0, 2i/LIF-cultured mESCs were dissociated and then cultured as embryoid bodies with rotary suspension in DMEM/FBS medium. At Day 2.5, the embryoid bodies were plated onto gelatin in basal N2B27 medium, with either no supplement (producing ectoderm), BMP4 (producing mesoderm), or Activin/EGF (producing endoderm and epiblast stem cells). Every 24 hours from Day 0 to Day 11, cell samples were collected with rapid dissociation followed by immediate paraformaldehyde fixation. Mass cytometry analyses was then performed to simultaneously measure the following markers at the single-cell level in samples from every timepoint and culture condition: pluripotency markers (Oct4, Nanog, Sox2, Lin28, SSEA1, Klf4, Sall4), germ layer and cell type markers (Nestin, TUJ1, Flk-1, PDGFR-Î±, Desmin, GATA4, FoxA2, Cdx2, ICAM-1, CD44, CD24, CCR9, CD45, CD41), epithelial vs. mesenchymal status (EpCAM, Vimentin), cell signaling and cell cycle status (phospho-Stat3, Î²-Catenin, c-Myc, p53, cleaved-Caspase 3, phospho-Rb, Ki67, IdU), global epigenetic status (acetyl-Lysine 9 Histone H3, pan-acetyl Histone H4). Global, time-resolved, FLOW-MAP analysis of these high-dimensional single-cell datasets reveals continuous cellular trajectories from Nanog-positive/CD54-positive â□□ground-stateâ□□ pluripotent stem cells to TUJ1-positive neurons (ectoderm), GATA4-positive/FoxA2-negative cardiomyocytes (mesoderm), and GATA4-positive/FoxA2-positive (endoderm) cell types, as well as an Oct4-positive/EpCAM-positive epiblast stem cell-like population. Molecular characterization of the intermediate cell types along these trajectories and analysis of their branching structure reveals two alternative paths for ectoderm formation, and similarity-based entanglement between early mesoderm and endoderm progenitors. In addition to *in vitro* pluripotent stem cell differentiation, this single-cell time course-based analysis strategy with the FLOW-MAP algorithm can be applied to other dynamic cell systems *in vitro* and *in vivo*, including direct reprogramming, embryonic development, immune response, oncogenesis, and drug resistance.

Single-Cell Dissection of Muscle Stem Cell Functional Heterogeneity

5th International Conference on Stem Cell Engineering

- Conference: International Conference on Stem Cell Engineering
- Year: 2016
- Proceeding: International Conference on Stem Cell Engineering 2016
- Group: General Submissions
- Session:

  Computational Stem Cell Biology

- Time:

  Monday, October 24, 2016 - 12:00pm-12:15pm

Authors:
Cosgrove, B. D., Cornell University
Munson, B. P., Cornell University
Soueid-Baumgarten, S. Y., Cornell University
Chen, F. M., Cornell University

Muscle stem cells (MuSCs), also known as satellite cells, play a central role in maintaining muscle homeostasis and are regulated by a highly choreographed cascade of cell-fate changes during muscle regeneration. MuSC self-renewal function progressive declines in advanced aging, resulting in a loss of regenerative capacity and muscle atrophy. This functional decline is caused by both muscle microenvironment alterations and dysregulation of key stem cell-intrinsic regulators, together leading to a restricted self-renewal function, which manifests heterogeneously, of MuSCs in aged tissues. We have previously linked MuSC functional variation with aging to aberrant p38 MAPK pathway regulation, but these alterations have not be connected to changes in the MuSC receptor repertoire.

Here, we used single-cell RNA-sequencing of FACS-sorted MuSCs from the hindlimbs of uninjured and notexin-injured young (3 months-old), aged (20 mo), and geriatric (26 mo) mice to gain an unbiased picture of differences in gene expression within individual cells and populations during the aging process. We optimized a Smart-Seq2 protocol to generate single-cell RNA-sequencing libraries with minimal read bias and high yield after sequencing on a NextSeq instrument. We analyzed the single-cell transcriptomes using Monocle to generate hierarchically organized maps of MuSC sub-population variation and diversification. From these, we observed that a family of non-ligand binding transmembrane proteins exhibited significantly heterogeneous single-cell expression patterns in aged and injured MuSCs.

In injury-response experiments, we confirmed this expression diversification at the protein level within the comprehensive Î±7-integrin$^+$ CD34$^+$ MuSC population by FACS immunostaining. We found that a particular member of this protein family is expressed trimodally in injured and aged MuSCs, but unimodally in uninjured MuSCs from healthy young adult mice. In particular, we found that geriatric muscles (26 mo) had increased frequencies of both low and high-expressing subpopulations, indicating that this trifurcated expression pattern might distinguish functional from dysfunctional MuSCs.

To experimentally test the self-renewal function of these MuSC sub-populations, we isolated cells from transgenic *Luciferase* donor mice by FACS, fractionated them into low/mid/high-expressing groups, and transplanted 50 cells from each group into hindlimb muscles of irradiated *NOD/scid* recipient mice. We measured donor cell-mediated muscle repair by bioluminescence imaging and immunohistochemistry at one-month post-transplant. We observed that only the â□□midâ□□ subpopulation can engraft and repair muscles, indicating that the â□□lowâ□□ and â□□highâ□□ subpopulations are defective in self-renewal *in vivo*.

To assess the molecular changes that may explain these functional discrepancies, we performed RNA-sequencing on pools of these three MuSCs subpopulations from young mice. We observed that the receptor-mid population was enriched for a Pax7^{high} / Myogenin^{low} expression pattern typical of quiescent MuSCs and the â□□highâ□□ and â□□lowâ□□ subsets cells had a Pax7^{low} / Myf5^{high} pattern typical of myogenic progenitors.

This approach demonstrates the utility of single-cell RNA-sequencing in the unbiased identification of stem-cell subpopulation functional diversification in aging. Our findings suggest specific MuSC subpopulations that may represent the dysfunctional stem cells in aged muscles, and provide new molecular candidates for enhancing MuSC function in the elderly.

Engineering Organoid Development in 3D Artificial Microenvironments

5th International Conference on Stem Cell Engineering

- Conference: International Conference on Stem Cell Engineering
- Year: 2016
- Proceeding: International Conference on Stem Cell Engineering 2016
- Group: General Submissions
- Session:

  Microtechnologies & High Throughput Screening/Disease Model

- Time:

  Monday, October 24, 2016 - 1:45pm-2:15pm

Authors:
Lutolf, M., EPFL

The earliest steps of development are characterized by cellular reorganization and differentiation within a three-dimensional (3D) microenvironment. This 3D context allows for a complex spatial interplay between biochemical and physical signals, and governs important cellular rearrangements leading to morphogenesis. In vitro approaches have attempted to recapitulate key features of these processes, and it has now become possible to generate an increasing variety of self-organizing multicellular tissue constructs termed â□□organoidsâ□□. While important aspects of the 3D in vivo organization have been recreated in these organoid systems, such studies have been exclusively performed in Matrigel™, a poorly defined 3D microenvironment whose properties cannot be readily modulated. As such, the uncharacterized interactions between cells and this extracellular matrix (ECM) have proven to be a major challenge to understanding the underlying regulatory mechanisms governing morphogenesis. In this talk, I will highlight recent efforts in my lab to employ tunable synthetic hydrogels in order to disentangle the contributions of biochemical and physical components of the microenvironment in the specification of stem cell fate and morphogenesis.

Heart-on-a-Plate for Modelling of Cardiac Physiology and Drug Screening

5th International Conference on Stem Cell Engineering

- Conference: International Conference on Stem Cell Engineering
- Year: 2016
- Proceeding: International Conference on Stem Cell Engineering 2016
- Group: General Submissions
- Session:

  Microtechnologies & High Throughput Screening/Disease Model

- Time:

  Monday, October 24, 2016 - 2:15pm-2:30pm

Authors:
Radisic, M., University of Toronto

The current drug development process is both very slow (15 year average) and costly ($1.5B/drug average). Despite this hefty investment, inefficiencies in the drug screening process routinely result in the withdrawal of drugs from the market due to serious toxicities and adverse cardiovascular effects. Safety screen assessments performed on cell cultures and animal models do not always correlate with clinical risk. Specifically, this entire framework predicts only 75-90% of the effects of a drug in a human patient, resulting in the abandonment of many promising drugs early in development and enabling some cardiotoxic drugs go to market. A potential explanation is that much of the safety and efficacy testing has relied upon animal studies and little information has been gathered from human preparations. The applicability of animal studies is limited by the fundamental cardiovascular differences amid species. Similarly, cardiac ion channel *in vitro* screens frequently involve non-cardiac cells (Chinese hamster ovary, CHO or human embryonic kidney, HEK) that have been engineered to overexpress an individual target ion channel, most commonly the human ether-a-go-go (hERG) channel (I_{Kr} current). However, interference with other currents may also contribute to cardiac complications necessitating an assessment protocol that evaluates the effect of a drug on the collective behaviour of cardiac ion channels.

Organ-on-a-chip devices hold a promise to revolutionize the way drugs are discovered and tested when used in conjunction with cardiomyocytes derived from human pluripotent stem cells. Current devices are limited by the presence of non-physiological materials such as glass and drug-absorbing PDMS as well as the necessity for specialized equipment such as vacuum lines and fluid pumps that inherently limit their throughput. An overview of two new technologies, Biowire and AngioChip, that overcome some of the limitations above will be presented. Biowires and AngioChips are situated in an open-concept plate-like platform compatible with current liquid handling practices that involve pipetting or pipetting robots. The use of these technologies in maturation of cardiomyocytes derived from pluripotent stem cells, their perfusion, drug testing and disease modelling will be presented.

*** Formatting Issue - Best Available Paper/Graphic ***

Microtopography-Enhancement on Dopaminergic Neuronal Differentiation

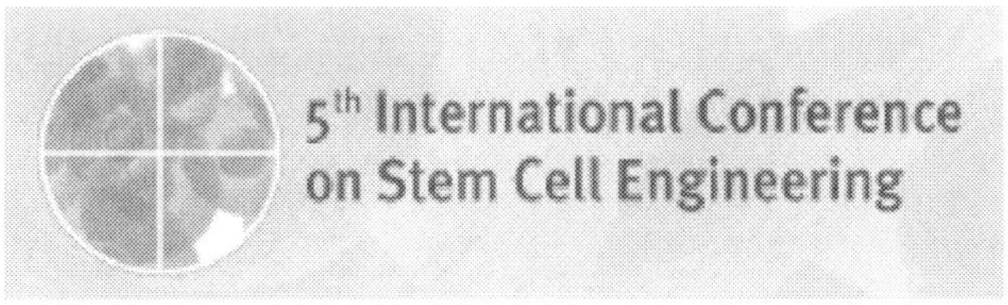

- Conference: International Conference on Stem Cell Engineering
- Year: 2016
- Proceeding: International Conference on Stem Cell Engineering 2016
- Group: General Submissions
- Session:

  Microtechnologies & High Throughput Screening/Disease Model

- Time:

  Monday, October 24, 2016 - 2:30pm-2:45pm

Authors:
Yim, E., University of Waterloo
Tan, K.
Chai, C.
Lim, K. L., Duke-NUS Graduate Medical School
Goh, E., National Neuroscience Institute
Lim, W. M.

Micro and nanotopograhy have been shown to be able direct stem cell fate. Substrate topography also has the potential in differentiating stem cells into subtype specific cells. Parkinsonâ□□s disease is a neurodegenerative disease attributed to the loss of midbrain dopaminergic (DA) neurons. Differentiation of stem cells into subtype specific cells may be guided by appropriate topographical cues but the role of topography has hitherto not been well understood. We aim to develop an in vitro substrate that will accelerate the derivation of midbrain dopaminergic neurons from human pluripotent stem cells (hPSCs) and neural progenitor cells.

Neuronal differentiation of murine hippocampal neural progenitor cells (mNPCs) on a Multi-Architecture (MARC) chip with various topographical structures was performed to identify the topographies that generated the highest percentage of neuronal (b-III-tubulin positive) and dopaminergic (tyrosine hydroxylase positive) populations. Subsequently, the ability of topographical patterns to influence the differentiation of hPSCs into subtype-specific and regionalized dopaminergic neurons was investigated. We have made minor modifications to the protocol based on dual SMAD inhibition method and optimized on patterned substrates to further improve the efficiency for dopaminergic neuron derivation.

From both type of stem cells, the topographical patterns could enhance the dopaminergic neuron differentiation. These results show the use of topographical influence for neuronal subtype specification, which could be translated into new therapeutic approaches for this neurodegenerative disease.

*** Formatting Issue - Best Available Paper/Graphic ***

Modeling the Pharmacogenomics of Chemotherapy-Induced Cardiotoxicty Using hiPSC

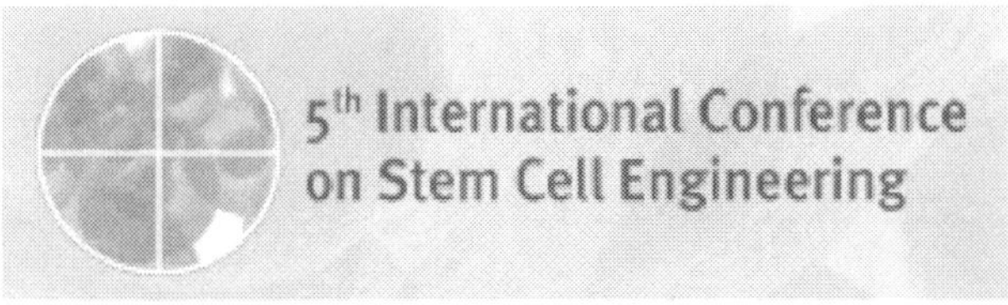

- Conference: International Conference on Stem Cell Engineering
- Year: 2016
- Proceeding: International Conference on Stem Cell Engineering 2016
- Group: General Submissions
- Session:

  Microtechnologies & High Throughput Screening/Disease Model

- Time:

  Monday, October 24, 2016 - 2:45pm-3:00pm

Authors:
Burridge, P., Stanford University

It is well established that numerous chemotherapy drugs, whilst effective in treating a wide range of malignancies, can lead toÂ cardiotoxicity that can lead to dysrythmias, cardiomyopathy, and heart failure.

At present, it is not possible to predict which patients will be affected by chemotherapy-induced cardiotoxicity (CIC). Here we demonstrate that patient-specific human induced pluripotent stem cellâ□□derived cardiomyocytes (hiPSC-CMs) can recapitulate the predilection to CIC of individual patients at the cellular level. hiPSC-CMs derived from individuals with cancer who experienced CICÂ were consistently more sensitive to chemotherapy toxicity than hiPSC-CMs from patients who did not experience CIC. Our data indicate that hiPSC-CMs are a suitable platform to identify and characterize the genetic basis and molecular mechanisms of CIC.Â

Isolation and Characterization of Lung Cancer Stem Cells from p53 Gain-of-Function Mutant Mouse Models

5th International Conference on Stem Cell Engineering

- Conference: International Conference on Stem Cell Engineering
- Year: 2016
- Proceeding: International Conference on Stem Cell Engineering 2016
- Group: General Submissions
- Session:

  Microtechnologies & High Throughput Screening/Disease Model

- Time:

  Monday, October 24, 2016 - 3:00pm-3:15pm

Authors:
Rao, R.
Gadepalli, V.

Recent advances in understanding lung biology have shown evidence for the existence of resident lung stem cells. Independent studies in identifying and characterizing these somatic lung stem cells have shown the potential role of these cells in lung repair and regeneration. While identifying a homogenous stem cell pool with unique cell surface markers is still a challenge, the currently available data support the existence of different lung cells with stem cell characteristics of self-renewal and differentiation. Understanding the functional characteristics of these tissue resident stem/progenitor cells has gained much importance with increasing evidence of cancer stem cells, cells in a tumor tissue with stem cell characteristics. Lung cancer is most commonly characterized by loss of p53 function which results in uncontrolled cell divisions. Incidence of p53 point mutations is highest in lung cancer, with a high percentage of missense mutations as a result of tobacco smoking. Certain point mutations in p53 gene results in its oncogenic gain of functions (GOF), with enhanced tumorigenic characteristics. However, there are no available data on characterization of lung stem cells carrying GOF mutations and correlating them with those of normal stem cells. In this study, for the first time we characterized lung stem/ progenitor cells from cancer sensitized mice carrying p53 GOF mutations. Our results show that there is distinct morphological and cell proliferation difference between the Sca-1 expressing cells established from mice lungs carrying different p53 genotypes. Additionally, our results suggest that Sca-1 expressing subpopulation of cells are significantly higher in p53 mutant carrying mice lungs than those in $p53^{+/+}$ wild type. We also noted that Sca-1 expressing p53 mutants show tumorigenic properties of p53 accumulation, migration, and colony formation in non-adherent culture conditions. When the cells across the different genetic backgrounds were subjected to mesenchymal lineage differentiation, we noted that $Sca\text{-}1^{High}$ $p53^{+/+}$ wild type readily differentiated into adipogenic and osteogenic lineages. However, the mesenchymal differentiation ability of the cells isolated from p53 mutants is not evident compared to their wild type counterparts. Results from our studies demonstrate the stem cell and tumorigenic nature of Sca-1 cells carrying p53 GOF mutations. This study lays the foundation and rationale for elaborate cellular and molecular level investigation of resident stem/progenitor cells expressing p53 GOF mutations. The insights on these resident stem/progenitor cells will ultimately shed light on cancer progression and development of new models and therapeutics for lung cancer.

High-Efficiency Cellular Reprogramming and Differentiation with Microfluidics

5th International Conference on Stem Cell Engineering

- Conference: International Conference on Stem Cell Engineering
- Year: 2016
- Proceeding: International Conference on Stem Cell Engineering 2016
- Group: General Submissions
- Session:

  Microtechnologies & High Throughput Screening/Disease Model

- Time:

  Monday, October 24, 2016 - 3:15pm-3:30pm

Authors:
Elvassore, N., University of Padova

Drug development is a time-consuming process with high failure rate. Moreover, even multiple approved drugs show patient-to-patient variability as for efficacy and toxicity response.

We report that the efficiency of reprogramming human somatic cells to induced pluripotent stem cells (hiPSCs) can be dramatically improved in a microfluidic environment. microliter-volume confined environment resulted in a 50-fold increase in efficiency over traditional reprogramming by delivery of synthetic modified mRNAs encoding transcription factors. We found that in these small volumes, extracellular components of the TGF-b and other signaling pathways exhibited temporal regulation that appears critical to acquisition of pluripotency. The high quality and purity of the resulting hiPSCs allowed direct differentiation into functional hepatocyte- and cardiomyocyte- like cells in the same platform without additional expansion.

These results is step forward the high-throughput generation of patient-specific human pluripotent stem cells.

Dental Pulp Stem Cell Adhesion and Differentiation on a Novel Therapeutic Biomaterial

5th International Conference on Stem Cell Engineering

- Conference: International Conference on Stem Cell Engineering
- Year: 2016
- Proceeding: International Conference on Stem Cell Engineering 2016
- Group: General Submissions
- Session:

  In situ/Endogenous Regeneration

- Time:

  Monday, October 24, 2016 - 12:00am-12:01am

Authors:
Scherba, J. C., Wyss Institute for Biologically Inspired Engineering
Mooney, D. J., Harvard University
Vining, K. H., Harvard University
Celiz, A. D., University of Nottingham

Dental pulp stem cell adhesion and differentiation on a novel therapeutic biomaterial

Jacob C. Scherba[1,2], Kyle H. Vining[1,2], Adam D. Celiz[1,3], David J. Mooney[1,2].
1. Wyss Institute for Biologically Inspired Engineering. 2. Harvard John A. Paulson School of Engineering and
Applied Sciences. 3. University of Nottingham.

Introduction: Over 1 billion dental fillings are performed annually in the United States according to the U.S. Food and Drug Administration. Current dental decay treatments, however, can lead to pulp injury that may necessitate costly and painful procedures, like root canals and extractions. Using a high-throughput microarray, we identified AT03 as a leading candidate to facilitate dental pulp stem cell (DPSC) adhesion. Subsequently, we scaled to a bulk biomaterial that can be cured *in situ* via UV irradiation. In this study, we seek to understand the mechanism of DPSC adhesion to our material and investigate its ability to allow for differentiation of DPSCs into odontoblasts. Understanding these processes is the basis for enabling dentin regeneration in teeth, and it informs our investigation of the therapeutic capability of our material *in vivo*. The mechanistic understanding of the behavior of these cells on our material, if applied properly, has considerable implications for routine therapeutic dentistry.

Materials and Methods: To investigate the mechanism of adhesion, RNA was isolated from adhered cells on AT03 and Bis-GMA (current industry standard) and tissue culture plastic (TCP) controls at 8 and 48 hours. F- actin and immunostaining were performed to illustrate cell morphology and focal adhesion as well as proliferative activity. For differentiation, RNA was collected from adhered DPSCs at 21 days. RT qPCR was performed to explore relative levels of odontoblastic and osteogenic differentiation.

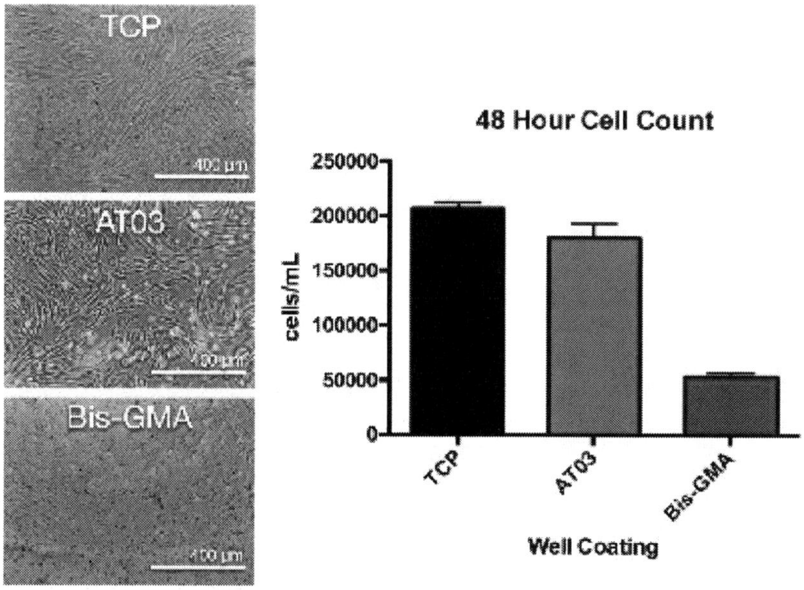

Results and Discussion: Phase microscopic images qualitatively determined that human DPSCs adhered more strongly to our AT03 polymer than to Bis-GMA. In addition, cell counts were performed after 48 hours to determine the number of viable cells. There were significantly fewer viable DPSCs on Bis-GMA compared to both TCP and our AT03 polymer ($p < 0.0005$). Gene expression analysis showed that AT03 supports expression of osteogenic differentiation markers after 21 days in culture.

****

Figure 1. Adhesion of DPSCs is illustrated in the phase microscope images to the left, characterized by the distinct elongated morphology of the cell. In addition, a significant reduction in viable cell count is exhibited

between Bis-GMA and both AT03 and TCP. Data is presented + SD.

Conclusions: Human DPSCs adhere more strongly to our AT03 polymer than to the industry standard polymer, Bis-GMA, and there are more viable cells present on our material over time. The fact that our material allows for the growth of viable DPSCs could radically alter therapeutic dentistry. The facilitation of osteogenic differentiation is also promising because hard tissue mineralization plays a key role in dental tissue repair. Applied in the clinic, our material could allow for the regeneration of dentin in patients who might otherwise require expensive and painful procedures like extractions or root canals. We are currently investigating the behavior of our material *in vivo* using a rat dental pulp injury model.

*** Formatting Issue - Best Available Paper/Graphic ***

Development of an Improved Vascular Graft for Use in Children

5th International Conference on Stem Cell Engineering

- Conference: International Conference on Stem Cell Engineering
- Year: 2016
- Proceeding: International Conference on Stem Cell Engineering 2016
- Group: General Submissions
- Session:

  In situ/Endogenous Regeneration

- Time:

  Monday, October 24, 2016 - 4:00pm-4:30pm

Authors:
Breuer, C.

Despite significant advances in the care of infants and children born with congenital heart disease, congenital cardiac anomalies represent a leading cause of death in the newborn period. One significant source of morbidity and mortality arises from the use of man-made synthetic materials in the form of vascular patches, grafts, and replacement heart valves in most operations. Use of synthetic materials are associated with an increased risk of thromboembolic complications, increased susceptibility to infection, and perhaps most importantly lack of growth potential. We developed a tissue engineered vascular graft created by seeding an individualâ□□s own cells onto a biodegradable tubular scaffold in order to create an autologous, living vascular conduit with the ability to grow, repair, and remodel. We subsequently translated this technology from the bench to the clinic performing the first clinical trial evaluating the use of tissue engineered vascular grafts in humans. Results of this trial confirmed the growth capacity of the tissue engineered vascular graft making it uniquely suited for use in children but also revealed that stenosis was the primary graft related complication effecting over 25% of graft recipients. We subsequently performed a series of investigations in order to elucidate the molecular mechanisms underlying neotissue formation and the stenosis. Based on our studies we discovered that the seeded cells do not directly contribute to vascular neotissue but instead modulate neotissue formation via a paracrine mechanism. We also discovered that vascular neotissue arises from the neighboring vascular tissue through an immune mediated regenerative process and that excessive inflammation causes stenosis. Next based on our mechanistic discoveries we rationally designed strategies for inhibiting the formation of tissue engineered vascular graft stenosis by altering the foreign body reaction caused by the tissue engineering scaffold. Most recently we are working translate our most recent work to the clinic through the development of the second generation tissue engineered vascular graft.

Exosomes (EXOs) Derived from Progenitors Renal CELL (RPCs) Minimized the ACUTE Kidney Injury (AKI) By LPS MODEL

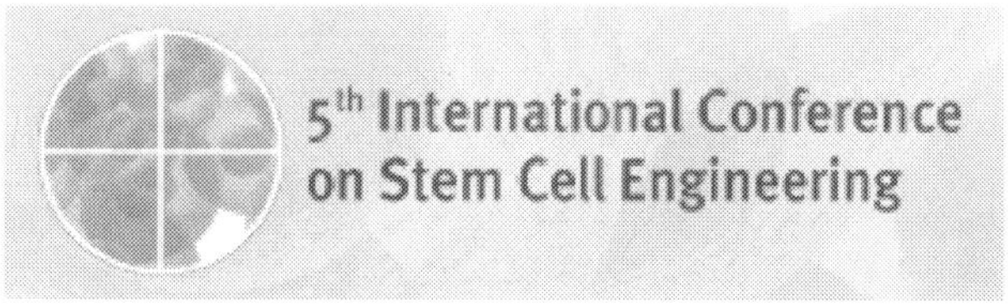

- Conference: International Conference on Stem Cell Engineering
- Year: 2016
- Proceeding: International Conference on Stem Cell Engineering 2016
- Group: General Submissions
- Session:

  In situ/Endogenous Regeneration

- Time:

  Monday, October 24, 2016 - 4:30pm-4:45pm

Authors:
Reis, L. A.
Borges, F. T., UNIFESP/EPM
DA Gloria, M. A., UNIFESP/EPM
Sinigaglia-Coimbra, R. D. C., UNIFESP/EPM
Simoes, M. D. J., UNIFESP/EPM
Schor, N.

Introduction and Aim: AKI is followed by regeneration of damaged renal tubular epithelial cells. This damage is caused by toxic and ischemic insults to the kidney; it is characterized by acute tubular necrosis (ATN). There are possible sources of regenerative tubular cells as bone marrow that home to the injured kidney or resident renal stem cells as RPCs. The aim of this study was characterize RPCs and the effect of EXOs derived these cells in AKI induced by LPS. **Method:** The cells were obtained from neonate kidney (1st born day) and characterized by fluorescence and confocal microscopy (CD24, PAX2 or CD133); the EXOs by transmission electronic microscopy (TEM) and fluorescence (CD63). The RPCs were preconditioned (PC) with LPS (10µg/ml) or hypoxia (12 hours) and then, the EXOs were quantified using Nanosight and Western blot. Then, the LLCP-K1 cells were treated (72hours) with LPS, PBS (CTL) or

LPS+EXOs-PC and cellular viability (%), NO production (nmoles/mg protein) were evaluated by Trypan blue and Griess method, respectively. The mices (C57BL6J) received PBS (vehicle of LPS) or LPS 10mg/BW i.p. (one administration), after 24 hours from the insult, the animals received EXOs (100μg/ml) into caudal vein. After 72 hours of PBS or LPS administration, the animals were killed and the kidneys were submitted to HE (μm^2), NGAL, IL18 or KIM 1 (%) quantifications and creatinine (Cr) and urea (U) evaluations (mg/dl). **Results:** The cells were positive for PAX2, CD24 and CD133 the EXOs were positive for CD63 and measured less than 100μm diameter. LLCP-K1 treated with LPS showed lower cellular viability when compared to CTL (79±12.6 *vs.*99±14.9 %; $p<0.05$) and higher in LPS+EXOs-PC with LPS or hypoxia (92±14.3 and 93.4±7.8, respectively, $p<0.05$). NO production were higher in LPS vs. CTL (189±5.6 *vs.*76±8.2; nmoles/mg protein, $p<0.05$) and lower in LPS+EXOs PC with LPS or hypoxia groups (113±11.4 and 128±18.2, $p<0.05$). The animals treated with LPS showed increase in Cr and U when compared to CTL group (0.34±0.02 *vs.*0.8±0.03; 69±3.4 *vs.*12±1.9 mg/dl; $p<0.05$; respectively). When the animals were treated with EXOs (LPS+EXOs group), the Cr and U were lower compared with LPS group (0.54±0.03 *vs.*0.8±0.03; 24.8±x vs. 69±3.4 mg/dl, $p<0.05$, respectively). We observed higher ATN (%) and labeling intensity of renal biomarkers as NGAL, KIM1 and IL18 (%) in LPS when compared to CTL, respectively (14±2.1 *vs.* 6.4±0.8; 26.7±5.7 *vs.*1±0.07; 37.3±6.2 *vs.*0.5; 9.9±2.3 *vs.*0.8±0.03; $p<0.05$). These parameters evaluated were minimized when the animals received EXOs (8.0±0.9; 11±2.1; 11.3±0.9; 6.9±0.07; p<o.o5; respectively). **Conclusion:** The treatment of AKI with EXOs derived by RPCs significantly ameliorated the ATN, functional kidney damage and renal biomarkers. When these cells were PC with LPS or hypoxia, the cellular viability were higher and NO production lower. The ability of these cells to repair renal damage are present in kidney and are activated when the kidney damage is occurring, suggesting their potential in the treatment of AKI.

Funds from: FAPESP, CNPq, CAPES and FOR

*** Formatting Issue - Best Available Paper/Graphic ***

Endogenous Neural Stem Cell Stimulation and Motor Recovery after Local Delivery of Cyclosporine and Erythropoietin with a Hyaluronan-Based Hydrogel

5th International Conference on Stem Cell Engineering

- Conference: International Conference on Stem Cell Engineering
- Year: 2016
- Proceeding: International Conference on Stem Cell Engineering 2016
- Group: General Submissions
- Session:

  In situ/Endogenous Regeneration

- Time:

  Monday, October 24, 2016 - 4:45pm-5:00pm

Authors:
Tuladhar, A., University of Toronto
Morshead, C., University of Toronto
Obermeyer, J., University of Toronto
Payne, S., University of Toronto
Shoichet, M. S., University of Toronto
Zand, S.

Endogenous neural stem cell stimulation and motor recovery afterlocal delivery of cyclosporine and erythropoietin with a hyaluronan-based hydrogel

Anup Tuladhar[1], Jaclyn Obermeyer[1,3], Samantha Payne[1,3], Sohrab Zand[5], Cindi M Morshead[1,2,6], Molly S Shoichet[1,3,4,6]

1: Institute of Biomaterials and Biomedical Engineering, 2: Department of Surgery, 3: Department of Chemical Engineering and Applied Chemistry, 4: Department of Chemistry, 5:Department of Human Biology 6:Donnelly Centre for Cellular and Biomolecular Research, University of Toronto, Toronto, ON, Canada

Introduction: Neural stem and progenitor cells (NSPCs) found in the adult brain hold promise for neural repair after brain injuries such as stroke. It has been been shown that drugs can be used to stimulate endogenous NSPCs to promote self-repair of the injured brain. Two such drugs are the small molecule immunosuppressant cyclosporine (CsA) and the growth hormone erythropoietin (EPO). In animal models of stroke injury these compounds have been individually found to increase NSPC survival, improve tissue repair and promote functional recovery (CsA); and to improve neurogenesis and angiogenesis (EPO). However, clinical use of these drugs for neurorepair requires overcoming the blood-brain barrier, a specialized structure in the neurovasculature that hinders drug accumulation in the brain after systemic drug delivery.

Local drug delivery can circumvent the blood-brain barrier, avoiding complications due to systemic side effects and dose-limiting toxicities. Unfortunately, the current clinical method causes further tissue damage due to cannula insertion into brain tissue and has an increased risk of infection. Our lab has developed a bioactive hydrogel for local drug delivery. The gel, HAMC, is a physical blend of hyaluronan and methylcellulose. HAMC can be implanted on the brainâ□□s surface without damaging tissue. By mixing drug-loaded polymeric particles into the gel we can locally deliver drugs for several weeks. Herein we investigate local co-delivery of CsA and EPO to stimulate motor recovery and tissue regeneration in stroke-injured rats.

Methods: HAMC hydrogel was mixed with CsA-loaded and EPO-loaded poly(lactic-co-glycolic acid) (PLGA) microparticles and injected onto the ispilateral brain surface of rats with an endothelin-1 stroke. We measured CsA and EPO diffusion over time in dorsoventral brain sections using mass spectrometry

and ELISA. Next, we evaluated behavioral recovery in rats treated with drug-loaded HAMC starting at 4 days after stroke. The combination of CsA and EPO was compared to CsA only or EPO only, and the appropriate controls, using 3 assays for sensorimotor function: the Montoya staircase task, forepaw asymmetry task and tapered beam traversing task.

Results: Local delivery with HAMC successfully provided sustained CsA and EPO release to the brain for 28 days post-implant. Drug accumulation in the subcortical NSPC niche was found to be greater than or equal to concentrations found in the brain after systemic administration (CsA) or endogenously (EPO). In the recovery of skilled forepaw use, measured with the Montoya staircase task, only animals treated with the combination with CsA and EPO showed significant recovery compared to untreated controls at day 10 and day 20 post- injury ($p<0.05$, two-way ANOVA with Dunnettâ□□s post-hoc test), whereas animals treated with either drug alone or with the vehicle (HAMC with blank PLGA microparticles) did not. Data analysis of the remaining sensorimotor assays is on-going.

Conclusions: HAMC hydrogel enabled sustained local delivery of CsA and EPO to the rat brain that resulted in motor recovery after stroke, demonstrating for the first time a synergistic effect of CsA and EPO on motor recovery. On-going work is evaluating endogenous NSPC stimulation and tissue regeneration after CsA and EPO delivery with HAMC.

Acknowledgements: The authors are grateful to the Canadian Institute for Health Research, Ontario Graduate Scholarship, Natural Sciences and Engineering Research Council, Training Program in Regenerative Medicine, Heart and Stroke Foundation and Canadian Partnership for Stroke Recovery for funding.

The Role of Progenitor Cells in Extracellular Matrix Scaffold Remodeling

5th International Conference on Stem Cell Engineering

- Conference: International Conference on Stem Cell Engineering
- Year: 2016
- Proceeding: International Conference on Stem Cell Engineering 2016
- Group: General Submissions
- Session:

  In situ/Endogenous Regeneration

- Time:

  Monday, October 24, 2016 - 5:00pm-5:15pm

Authors:
Gilbert, T., University of Pittsburgh

Extracellular matrix (ECM) scaffolds derived from various species and tissue sources have been investigated for use in the field of regenerative medicine, and many are routinely used in clinical practice. The host response to the ECM scaffold depends on both the origin of the source tissue and the processing steps utilized to prepare the scaffold for use. For a subset of ECM scaffolds, the host response has been characterized as constructive remodeling, which has been defined as the deposition of site-appropriate tissue at the site of scaffold placement with concomitant resorption of the scaffold. The timing of the resorption of the scaffold is an important determinant of whether constructive remodeling can occur, as the host response to the wound milieu is different when the scaffold is being enzymatically degraded as opposed to being integrated. The host response associated with constructive remodeling includes the presence of an M2 macrophage response and the presence of local and circulating progenitor cells. The talk will summarize the data available on this topic.

Electrical Function of Human Pluripotent Stem Cell-Derived Cardiomyocyte Grafts

5th International Conference on Stem Cell Engineering

- Conference: International Conference on Stem Cell Engineering
- Year: 2016
- Proceeding: International Conference on Stem Cell Engineering 2016
- Group: General Submissions
- Session:

  Differentiation & Morphogenesis II

- Time:

  Tuesday, October 25, 2016 - 8:30am-9:00am

Authors:
Laflamme, M. A., University Health Network

Human pluripotent stem cells (hPSCs) have a number of potential advantages for use in myocardial infarct repair, including a tremendous capacity for expansion in the undifferentiated state followed by the ability to differentiate into phenotypically unambiguous cardiomyocytes. Our group has contributed to the development of efficient, reliable protocols to generate large quantities of hPSC-derived cardiomyocytes (hPSC-CMs), and we have shown that the transplantation of hPSC-CMs results in the partial remuscularization of injured hearts in multiple preclinical models. Despite this, a number of important challenges remain, including concerns about the immature and heterogeneous electrophysiological phenotype of hESC-derived cardiomyocytes, the ability of these cells to undergo appropriate electromechanical integration following transplantation, and the risk of graft-related arrhythmogenesis. In this presentation, I will describe our group's recent efforts to understand and improve the electrical function of hPSC-CM grafts in injured hearts, including 1) the development of improved protocols to promote the electrophysiological maturation of hPSC-CMs in vitro, 2) new optical mapping methods to study hPSC-CM electrical behavior in vivo, and 3) transplantation studies in a more relevant porcine infarct model.

Mechanical regulation of direct cell reprogramming

5th International Conference on Stem Cell Engineering

- Conference: International Conference on Stem Cell Engineering
- Year: 2016
- Proceeding: International Conference on Stem Cell Engineering 2016
- Group: General Submissions
- Session:

  Differentiation & Morphogenesis II

- Time:

  Tuesday, October 25, 2016 - 9:00am-9:15am

Authors:
Li, S., UC Berkeley - UCSF
Soto, J., UC Berkeley - UCSF
Wong, S. Y., UC Berkeley - UCSF

Direct conversion of differentiated cells into a completely different lineage has wide applications in regenerative medicine, disease modeling and drug screening. Although the roles of transcriptional factors and chemical compounds in direct reprogramming have been widely studied, the effects of extracellular matrix (ECM) and substrate stiffness on the reprogramming process are not well understood. Here we used induced neuronal reprogramming as a model system to investigate how the mechanical property of the ECM modulates this process. Transcriptional factors Brn2, Ascl1 and Myt1L were transduced into fibroblasts to generate induced neurons (iNs), but the efficiency was low, especially for adult cells. Culturing cells on soft substrates enhanced reprogramming efficiency. Surprisingly, we found a biphasic dependence with an intermediate stiffness promoting the most efficient direct conversion of fibroblasts into iNs, suggesting that the mechanical regulation of reprogramming process may have totally different mechanisms from stem cell differentiation. Interestingly, this stiffness effect was independent of the chemical components of ECM proteins. Further investigations suggest that several signaling pathways, cytoskeleton and nuclear matrix are involved in the mechanotransduction to cell reprogramming. Our findings provide insights into the roles of substrate stiffness in regulating direct reprogramming and the underlying mechanism, which will open a new avenue for the rational design of smart biomaterials for cell reprogramming.

*** Formatting Issue - Best Available Paper/Graphic ***

Biomaterials Strategies to Enhance the Therapeutic Potential of Human Neural Stem Cells for Neural Tissue Regeneration

5th International Conference on Stem Cell Engineering

- Conference: International Conference on Stem Cell Engineering
- Year: 2016
- Proceeding: International Conference on Stem Cell Engineering 2016
- Group: General Submissions
- Session:

  Differentiation & Morphogenesis II

- Time:

  Tuesday, October 25, 2016 - 9:15am-9:30am

Authors:
Li, X.

Transplantation of neural stem or progenitor cells (NSCs) represents a promising strategy to reconstruct the lesion cavity and promote brain tissue regeneration following traumatic brain injury. However, ongoing inflammation at the lesion site and the lack of supportive matrix structure and vasculature within the cavity present a hostile environment that results in low cell survival and poor control over diï¬□erentiation and engraftment of the transplanted cells. We optimized an in situ crosslinkable hyaluronic acid (HA) hydrogel conjugated with laminin-derived peptide as a cell delivery matrix. It generated a robust neovascular network within the hydrogel at the lesion cavity when injected at 3 days following the CCI injury in a rat TBI model, and promoted significant vasculature network formation filling the lesion site at 4 weeks to 6 months following hydrogel implantation. Using this tailored HA hydrogel, delivering hNSCs in the format of spheroids for brain regeneration was superior to single cell transplantation in promoting cell survival following implantation. The majority of the grafted cells diï¬□erentiated to Tuj1+ and MAP2+ neuronal progenitors and populated the entire lesion cavity. The human NSC-derived neuronal progenitors extended large numbers of axons into the host brain tissue. Since manipulation of cell transcriptional network is a more eï¬□ective approach to promote stem cell diï¬□erentiation, when compared to growth factors cocktails, we optimized a poly (Î²-amino ester) (PBAE)-based nanoparticle system to transfect hNSCs with a key transcription factors Neurogenin-2 (Ngn2) and generated a significantly larger number of neurofilament positive (NF+) mature neurons at the lesion site of CCI at 4 weeks post transplantation when compared to non-transfected cells. This approach holds great potential in promoting neuronal maturation in a tissue repair site, thus improving therapeutic outcomes of stem cell-based therapy for brain tissue regeneration.

Tracking and Predicting Reprogramming from a Biophysical Perspective

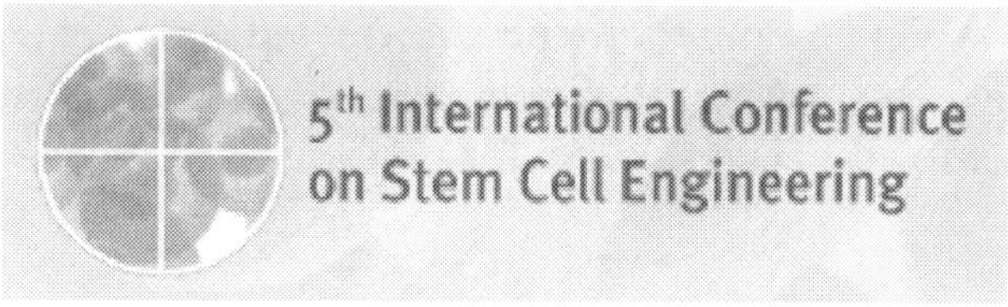

- Conference: International Conference on Stem Cell Engineering
- Year: 2016
- Proceeding: International Conference on Stem Cell Engineering 2016
- Group: General Submissions
- Session:

  Differentiation & Morphogenesis II

- Time:

  Tuesday, October 25, 2016 - 9:30am-9:45am

Authors:
Carlson-Stevermer, J., University of Wisconsin-Madison
Harkness, T.
Prestil, R., University of Wisconsin-Madison
Saha, K., Whitehead Institute for Biomedical Research
Seymour, S., University of Wisconsin-Madison

Molecular mechanisms of reprogramming terminally differentiated cells to a pluripotent state still are poorly understood. As a result, standard reprogramming techniques are noisy and inefficient. To address these shortcomings, we have developed a novel platform that allows for the dynamic tracking of subpopulations in a longitudinal manner across long periods of time. This platform is a simple approach that combines live-cell microscopy with surface-modified multiwell plates that separates thousands of cell populations. With this we are able to both watch and physically constrain cells into discrete islands during reprogramming. By watching subpopulations in real time we are able to distinguish intermediate states that either contribute or are detrimental to complete reprogramming. Furthermore by controlling the island geometry at the microscale, we are able to selectively activate mechanotransduction pathways (e.g., YAP/Taz) to promote transition through the endothelial-mesenchymal transition during the reprogramming process. Direct manipulation of the YAP/Taz pathway via lentiviral overexpression also affected the reprogramming process. Overall, this simple platform allows us to expand our understanding of intermediate cell states and increasing the overall efficiency of the reprogramming process.

Experimental and Computational Modeling for Directing Vascular Fate

5th International Conference on Stem Cell Engineering

- Conference: International Conference on Stem Cell Engineering
- Year: 2016
- Proceeding: International Conference on Stem Cell Engineering 2016
- Group: General Submissions
- Session:

  Differentiation & Morphogenesis II

- Time:

  Tuesday, October 25, 2016 - 9:45am-10:00am

Authors:
McCloskey, K., University of California, Merced

Experimental and Computational Modeling for Directing Vascular Fate

DE Glaser, WS Turner, L Wong, J Zamora, KE McCloskey, Ph.D.

University of California, Merced, CA

Embryonic stem cells (ESC) and induced pluripotent stem (iPS) cells and are attractive in vitro models of vascular development, therapeutic angiogenesis, and tissue engineering. Although a number of biochemical signals have been identified for directing endothelial fate, many of these factors activate redundant pathways, and the minimal combinatorial signals directing vascular fate have yet to be elucidated. Using our stage-specific chemically-defined derivation methodology, we examined multiple combinatorial factors for directing vascular smooth muscle cell (SMC) versus endothelial cell (EC) fate including: kinetics, cell seeding density, matrix signaling, matrix stiffness as well as medium treatment with vascular endothelial growth factor (VEGF), and basic fibroblast growth factor (bFGF). Combined with computational modeling, we have been able to identify the most potent interacting variables directing Flk-1+/KDR+ vascular progenitor cells (VPC), calponin+ SMC and VE-cad+ EC. The results indicate that temporal development in each of these stages is the most significant factor generating the desired cells. The generation of Flk-1+/KDR+ VPC from pluripotent ESC is directed predominantly by high cell seeding density and matrix signaling from fibronectin, while VEGF supplementation was shown to be superfluous. Additionally, stiffer materials directed more SMC while softer materials (10kPa) directed more EC. These results can now be applied towards the codifferentation of the two cells types in 3D materials for acquiring stable perfusable vasculature.

Engineering Personalized Neural Tissue Using Cell Permeable Transcription Factors

5th International Conference on Stem Cell Engineering

- Conference: International Conference on Stem Cell Engineering
- Year: 2016
- Proceeding: International Conference on Stem Cell Engineering 2016
- Group: General Submissions
- Session:

  Differentiation & Morphogenesis II

- Time:

  Tuesday, October 25, 2016 - 10:00am-10:15am

Authors:
Willerth, S., University of Victoria
Robinson, M.
Styan, T.
Vaidyanathan, R.

Introduction: The transcription factor Ascl1 plays a key role in neural development and previous work by the Wernig group demonstrated that Ascl1 overexpression using viral vectors can reprogram fibroblasts directly in neurons. Here we report on a recombinant version of the Ascl1 protein functionalized with intracellular protein delivery technology (Ascl1-IPTD) that can be used to rapidly differentiate human induced pluripotent stem cells (hiPSCs) into neurons. As hiPSCs can be generated from a patient's own cells, this method could potentially enable engineering of patient specific neural tissue.

Results: We evaluated a range of Ascl1-IPTD concentrations to determine the most effective amount for generating neurons from hiPSCs cultured in serum free media, which was 15 μg/mL. The protein did have concentration dependent effects on hiPSC differentiation into neurons. We then determined how often this protein needed to be added to the cell culture media to induce neuronal differentiation, selecting the following time points for media changes: 12, 24, 48, and 72 hours. These media changes were performed over 8 days and then the cells were stained for the presence of Ascl1-IPTD and the early neuronal marker Tuj1. All groups showed positive staining for the Ascl1 protein and exhibited high levels of Tuj staining, indicating that the frequency of media changes did not have an effect. Having demonstrated the effectiveness of the Ascl1-IPTD protein at differentiating hiPSCs into immature neurons, we then compared its ability to generate more mature neurons to a standard, commercially available protocol for generating neural progenitors and negative control cultures over a 12 day culture period. Both the cultures treated with Ascl1-IPTD and the positive control exhibited high levels of NeuN staining, indicating differentiation into mature neurons. Using Neurotrack software, we were able to analyze the average neurite length and amount of neurite branching observed for these cultures. The cells treated with Ascl1-IPTD had an average neurite length of 3.0372 mm/mm^2 compared to 2.1649 mm/mm^2 for the positive control. Likewise, the average number of branch points for the cells treated with Ascl1-IPTD was 28.052 1/mm compared to 16.743 1/mm for the positive control. The negative control showed little neurite extension and branching.

Discussion: While transcription factors play important roles in stem cell maintenance and differentiation, they are not often used to influence cell behavior due to their instability. The main goal of this work was to determine if using intracellular protein delivery technology could turn the transcription factor Ascl1 into a effective neuronal differentiator of stem cells, which it did. Our novel Ascl1-IPTD construct is both stable in the media and has a potent effect on stem cell behavior, making it an attractive alternative to viral based methods of increasing Ascl1 expression in cells. This study demonstrates how the Ascl1-IPTD protein can be used to efficiently and inexpensively derive neurons from hiPSCs compared to traditional protocols. Currently, we are determining if this Ascl1-IPTD protein can be used to directly reprogram human fibroblasts into neurons.

Biomaterials in Regenerative Engineering: immunoengineering and translation

5th International Conference on Stem Cell Engineering

- Conference: International Conference on Stem Cell Engineering
- Year: 2016
- Proceeding: International Conference on Stem Cell Engineering 2016
- Group: General Submissions
- Session:

  Immune Engineering

- Time:

  Tuesday, October 25, 2016 - 10:45am-11:15am

Authors:
Elisseeff, J.

Regenerative medicine and tissue engineering strategies aim to provide tissue substitutes for repair of defects due to disease, trauma, or congenital abnormalities. Biomaterials play a key role in tissue engineering strategies where they are designed to provide a three dimensional environment for cells, present biological cues, and implant in a physiological milieu. Defining biomaterial properties that have direct therapeutic relevance has inspired the focus on the immune response and its impact on not only biocompatibility but also regeneration directly. Translation of key technologies to clinical application in musculoskeletal and soft tissue reconstruction will also be presented.

Engineered Marrow Macrophages Engorge and Accumulate in Tumors before Differentiating into Tumor Associated Macrophages

5th International Conference on Stem Cell Engineering

- Conference: International Conference on Stem Cell Engineering
- Year: 2016
- Proceeding: International Conference on Stem Cell Engineering 2016
- Group: General Submissions
- Session:

  Immune Engineering

- Time:

  Tuesday, October 25, 2016 - 11:15am-11:30am

Authors:
Alvey, C., University of Pennsylvania
Title of Abstract

Engineered marrow macrophages engorge and accumulate in tumors before differentiating into TAMs

Cory Alvey,[1, 2] Kyle Spinler[1], Jerome Irianto[1], Charlotte Pfeifer[1], Yuntao Xia[1], Sangkyun Cho[1],

P.C.P Dave Dingal[1], Jake Hsu[1], Manu Tewari[1], and Dennis Discher,[1, 2*]

[1]Molecular and Cell Biophysics Laboratory, University of Pennsylvania, Philadelphia, PA, USA.

[2]Department of Pharmacology, Perelman School of Medicine, University of Pennsylvania, Philadelphia, PA, USA.

Abstract:

Monocytes and macrophages in cardiovascular and neuroinflammatory diseases are found to engorge and accumulate in solid tissues but mechanisms remain unclear. Within highly collagenous solid tumors, tumor associated macrophages (TAMs) are visibly motile, but limited studies also suggest TAMs are less phagocytic than macrophages in other tissues. Marrow-derived macrophages can be highly phagocytic, but whether they can also be made to accumulate in solid tumors and engulf cancer cells is questionable. By blocking marrow-derived macrophages from recognizing other cells as self, tail-vein injected donor cells accumulate in solid tumors in proportion to their engorgement on antibody-opsonized cancer cells. Blockade of SIRPA on both mouse and human donor macrophages inhibits binding of self-marker CD47 on targeted cells to drive tumor shrinkage for 1-2 wks, whereas TAMs can only slowly shrink opsonized tumors after tumor-CD47 knockdown. Nearly all donor macrophages in tumors unlike TAMs can be made to engulf tdTomato tumors, with phagocytosis impeding migration and causing accumulation, quantitatively predicting tumor shrinkage. However, expression profiles of donor macrophages change upon tumor-localization, with SIRPA increasing towards levels on non-phagocytic TAMs. RNA isolated from macrophages from tissues of varying stiffness shows that SIRPA expression correlates with microenvironment stiffness confirmed by *in vitro* studies using different stiffness gels. Nonetheless, multiple injections of engineered macrophages can re-initiate shrinkage of solid tumors *in vivo* during which blood parameters remain normal suggesting safety as well as efficacy.

Utilizing Humanized Mice to Study the Human Immune Response to Pro-Regenerative Biomaterials

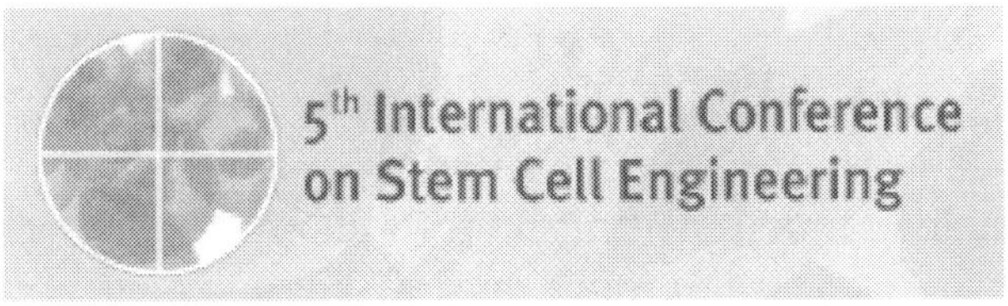

- Conference: International Conference on Stem Cell Engineering
- Year: 2016
- Proceeding: International Conference on Stem Cell Engineering 2016
- Group: General Submissions
- Session:

  Immune Engineering

- Time:

  Tuesday, October 25, 2016 - 11:30am-11:45am

Authors:
Wang, R., University of California, San Diego
Christman, K., University of California, San Diego
Johnson, T., Boston Consulting Group
Xu, Y., University of California, San Diego
He, J., University of California, San Diego
Rong, Z.

Introduction: The immune response plays an integral role in tissue repair and regulation of endogenous stem cells to regenerate damaged tissue. Effectiveness and safety of biomaterial therapies that stimulate repair are dependent on our understanding of their interaction with the human immune response. However, standard preclinical animal models provide limited representation of the human immune response and its numerous unique characteristics. We studied the use of a humanized mouse model, which generates multi-lineages of human immune cells, for evaluating the human immune response to xenogeneic and allogeneic decellularized extracellular matrix (ECM) based biomaterials.

Methods: NOD/SCID/γ immunodeficient mice were implanted with human liver, thymus and CD34+ fetal liver cells to generate humanized mice. Humanized and Balb/c mice were subcutaneously injected with myocardial ECM hydrogels derived from decellularized xenogeneic porcine and allogeneic human tissue, and a nondecellularized pro-inflammatory control. Injections isolated at 7 days were assessed for immune cell infiltration by immunohistochemistry and polarization by PCR (n = 8-12 injections). Immune cell infiltration was assessed overall and 200 m into the core of the injection. T-cells (CD3, CD4, CD8) and macrophages (F4/80, iNOS, CD206) were stained for specific phenotypic profiles. Pro-remodeling versus pro-inflammatory T-helper (Th-) cell (Th2/Th1) and macrophage (M2/M1) polarization was assessed by gene expression ratio of GATA3/TBX21 and ARG1/NOS2, respectively, with CRTH2/CCR5 uniquely assessed for human Th-cells. Results are reported as meanSEM with significance considered at $p < 0.05$.

Results: Although overall $CD3^{+}CD4^{+}$ Th-cell response to decellularized materials was not significantly different, significantly greater Th-cell infiltration into the material core was observed for porcine compared to human myocardial hydrogels (53.69.9 versus 14.36.9 cells/mm^2) in humanized mice with no significant difference in Balb/c infiltration (97.129.8 versus 92.110.4 cells/mm^2). Significantly fewer $CD3^{+}CD8^{+}$ cytotoxic T-cells and pro-inflammatory $F4/80^{+}iNOS^{+}CD206^{-}$ macrophages were found for the whole material in porcine (4.51.0, 44.58.7 cells/mm^2) and human (17.68.9, 31.811.5 cells/mm^2) compared to nondecellularized myocardial hydrogels (233.578.0, 129.2.7 cells/mm^2) in humanized mice with similar results observed in Balb/c mice. Numbers of pro-remodeling $F4/80^{+}iNOS^{-}CD206^{+}$ and triple stained $F4/80^{+}iNOS^{+}CD206^{+}$ macrophages were not significantly different for all materials in both mouse models, however, ratios of pro-remodeling $F4/80^{+}iNOS^{-}CD206^{+}$ versus pro-inflammatory $F4/80^{+}iNOS^{+}CD206^{-}$ macrophages suggested M2 trends for decellularized materials compared to nondecellularized. Gene expression ratio analysis (GATA3/TBX21, CRTH2/CCR5, ARG1/NOS2) demonstrated significant Th2 and trending M2 profiles for porcine (2.950.2, 2.760.3, 1.70.4) compared to nondecellularized (1.00.2, 1.00.6, 1.00.2) with Th2 trends for human myocardial hydrogels (1.360.2, 1.650.1, 0.840.2) in humanized mice. In comparison, Balb/c results (GATA3/TBX21, ARG1/NOS2) gave trending Th2 and significant M2 profiles for porcine (3.42.2, 3.31.3), and trends for human (2.10.5, 1.70.2) compared to nondecellularized myocardial hydrogels (1.00.4, 1.00.3).

Conclusions: Humanized mice uniquely responded to porcine myocardial hydrogels with greater Th-cell infiltration based on immunohistochemistry and a stronger Th2 and M2 profile based on gene expression ratios compared to human and nondecellularized materials, suggesting a greater pro-remodeling response to the xenogeneic material. These results demonstrate the use of this model for assessing the degree and polarization of the human immune response to biomaterial therapies, and could be used to further model endogenous regeneration.

Mechanical Programming of Mesenchymal Stromal Cells for Immunomodulation

5th International Conference on Stem Cell Engineering

- Conference: International Conference on Stem Cell Engineering
- Year: 2016
- Proceeding: International Conference on Stem Cell Engineering 2016
- Group: General Submissions
- Session:

  Immune Engineering

- Time:

  Tuesday, October 25, 2016 - 11:45am-12:00pm

Authors:
Shin, J. W., University of Illinois at Chicago College of Medicine

Mesenchymal stem/stromal cells (MSCs) have been tested in a number of clinical trials due to their anti-inflammatory efficacy even in the absence of major donor cell engraftment. This highlights the importance of better understanding how adoptively transferred MSCs respond to inflammatory environments and modulate the immune system in the host. While hydrogels have been used for prolonged donor cell delivery by physical protection from the host, it remains unclear whether mechanical parameters of hydrogels can be leveraged to biophysically control cytokine production from donor MSCs through mechanotransduction. We have begun to uncover relationships between different parameters of matrix mechanics and specific immunomodulatory mechanisms by MSCs. Matrix stiffness modulates the ability of MSCs to respond to TNFÎ± and recruit immune cells by regulating NF-kB activation. In contrast, the volume ratio between the matrix and the cell, not matrix stiffness, regulates the ability of MSCs to respond to IFNÎ³. In the absence of inflammatory stimulation, matrix stiffness influences constitutive secretion of some cytokines from MSCs and primes them for anti-inflammatory functions. The results thus provide a mechanistic basis for encapsulating MSCs in hydrogels with defined mechanical parameters to achieve maximal immunomodulatory efficacy. Controlling matrix mechanics at the single cell level will not only enable the translation of these insights into MSC therapy but also help reveal cell-to-cell heterogeneity in mechanical regulation of immunomodulation.

Bioprocess Engineering Strategies for Stem Cell-Based Therapies and Regenerative Medicine

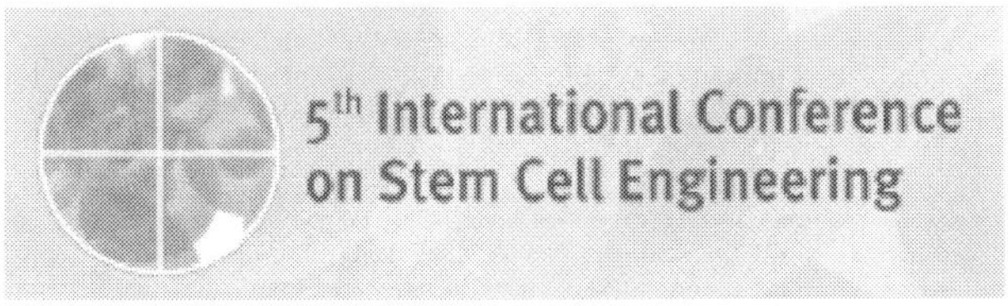

- Conference: International Conference on Stem Cell Engineering
- Year: 2016
- Proceeding: International Conference on Stem Cell Engineering 2016
- Group: General Submissions
- Session:

  Clinical Applications and Challenges (Including Scale Up and Imaging)

- Time:

  Tuesday, October 25, 2016 - 1:30pm-1:45pm

Authors:
Cabral, J. M. S., Instituto Superior Técnico, Universidade de Lisboa

Recent advances in the establishment of human embryonic, induced pluripotent and adult stem cell lines and the discovery of organ-specific stem cells with differentiation potential have substantially accelerated the potential of using these cells in regenerative medicine. Before any clinical application, stem cells need to be isolated, cultivated, differentiated into committed tissue precursor cells, stored and transported to the bed-side of a patient. The development of reliable in vitro systems for stem cell growth is a valuable tool to study the mechanisms controlling the expansion and differentiation of stem cells. Successful ex vivo models will enable the study of the dynamics and mechanisms of cell differentiation and organ development. The need for improved cultivation methods is also driven by the fact that most potential therapeutic applications are limited by the availability of stem cells or their derivatives. The in vitro propagation of stem cell populations remains undeveloped and is a major challenge because of the complex kinetics of the heterogeneous starting culture population, the transient nature of the subpopulations of interest and the complex interactions between the culture parameters. This lecture presents bioprocess concepts towards the ex vivo expansion and differentiation of stem cells in bioreactors, while maintaining their functional characteristics, including the ability to differentiate into appropriate receptive tissues.

Adult mesenchymal and pluripotent stem cells and their differentiation in neural and cardiac stem cells in microcarrier-based culture systems are analysed.

Patient-Derived Human Induced Pluripotent Stem Cells for Diabetes Cell Replacement Therapy and Drug Screening

5th International Conference on Stem Cell Engineering

- Conference: International Conference on Stem Cell Engineering
- Year: 2016
- Proceeding: International Conference on Stem Cell Engineering 2016
- Group: General Submissions
- Session:

  Clinical Applications and Challenges (Including Scale Up and Imaging)

- Time:

  Tuesday, October 25, 2016 - 1:45pm-2:00pm

Authors:
Millman, J. R., Massachusetts Institute of Technology

Patient-derived induced pluripotent stem (iPS) cells, differentiated to disease-relevant cells, are becoming quite important for the study of diseases due to their potential for cell replacement therapy and drug screening as well as improving our understanding of the disease pathophysiology. In Type 1 diabetes (T1D), pancreatic insulin-producing Î² cells are destroyed by an autoimmune attack and are not normally available for study, particularly without having undergone any disease-related stress. We have developed a strategy to generate stem cell-derived-Î² cells (SC-Î² cells) from iPS cells derived from T1D patients (T1D SC-Î² cells). This method utilizes three-dimensional cellular aggregates grown in suspension culture within spinner flasks and can produce up to half a billion cells per production run. These cells are able to function both in vitro and in vivo, have incredibly similar gene expression compared to isolated cadaveric Î² cells, and package their insulin into dense granules. When transplanted into mice that have had their own endogenous Î² cells destroyed with Alloxan, T1D SC-Î² cells can prevent diabetes and control blood glucose. Furthermore, we have developed an in vitro disease model of Î² cell stress using cytokine treatment, demonstrated partial rescue of this phenotype with an Alk5 inhibitor, and demonstrated that these cells respond to several different categories of anti-diabetic drugs. T1D SC-Î² cells have value in disease modeling and drug screening applications and, combined with autoimmune-modulating therapies, in cell replacement therapy.

*** Formatting Issue - Best Available Paper/Graphic ***

Microtissue Optimization for Practical Cell-Based Therapies

5th International Conference on Stem Cell Engineering

- Conference: International Conference on Stem Cell Engineering
- Year: 2016
- Proceeding: International Conference on Stem Cell Engineering 2016
- Group: General Submissions
- Session:

  Clinical Applications and Challenges (Including Scale Up and Imaging)

- Time:

  Tuesday, October 25, 2016 - 2:00pm-2:15pm

Authors:
Ungrin, M., University of Calgary
Toms, D.
Yu, Y.
Al-Ani, A., University of Calgary
Deardon, R.
Korbutt, G.

Substantial resources have been employed in establishing sources of transplantable material for cell-based therapies for various conditions. Many published protocols remain either in the proof-of-concept stage â□□ where the desired cell types can be generated, but in very limited numbers; or scalable, but with low overall efficiency. While illustrative of the tremendous potential of this area of research, in either of these cases significant challenges remain to be overcome in order to deliver a practical, economically viable cell-based therapy to the clinic.

Our research makes use of a microscale-tissue engineering platform technology which we developed, employing uniform size-controlled microtissues to provide a consistent cellular microenvironment. Building on this platform, we are establishing ways of systematically optimizing the production and / or assembly of cells to perform a given function. We were previously able to obtain a 36-fold increase in the yield of definitive endoderm cells from human pluripotent stem cells. I will present our current research, incorporating statistical Design of Experiments and Response Surface Methodology (DOE / RSM) into a generalizable approach for the optimization of cell production from pluripotent stem cells; and also the preparation of functional microtissues for therapeutic use. We have demonstrated the effectiveness of this approach both in vitro and in vivo, and propose that it should be considered by researchers envisaging future cell-based therapies as an outcome of their work.

Development of a Stem-Cell Derived Islet Replacement Therapy for Type 1 Diabetes

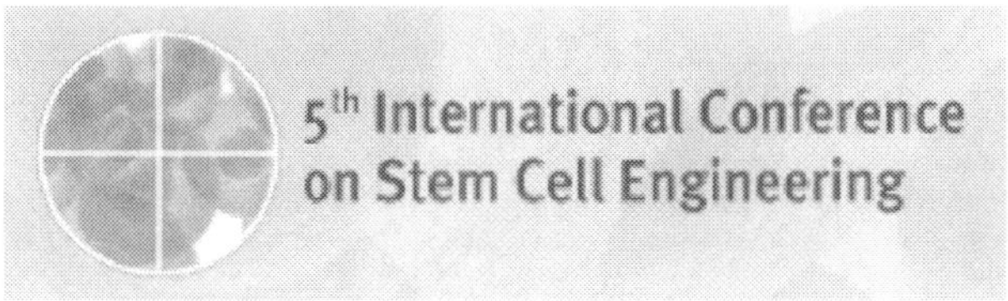

- Conference: International Conference on Stem Cell Engineering
- Year: 2016
- Proceeding: International Conference on Stem Cell Engineering 2016
- Group: General Submissions
- Session:

  Clinical Applications and Challenges (Including Scale Up and Imaging)

- Time:

  Tuesday, October 25, 2016 - 2:15pm-2:30pm

Authors:
D'Amour, K., Viacyte

ViaCyte Inc. is a clinical stage company developing a stem cell-based islet replacement therapy for treatment of patients with diabetes. The therapy is a combination product comprised of pancreatic endoderm cells encapsulated within a retrievable delivery device, ENCAPTRA® Drug Delivery System. After implantation, encapsulated progenitor cells differentiate into glucose-responsive, insulin-secreting cells. The renewable starting material for cell product manufacturing is human embryonic stem cells that are directed to differentiate to the pancreatic endoderm cell product using scalable processes. The bio-stable delivery device is designed to fully contain cells and to protect cells from immune attack, with the goal of eliminating the need for immunosuppressant drugs.

Funding in part from California Institute for Regenerative Medicine; SP1-06513, DR1-01423, TR1-01215 and JDRF.

*** Formatting Issue - Best Available Paper/Graphic ***

Imaging Stem Cell Delivery Near the Bed: Are We There Yet?

5th International Conference on Stem Cell Engineering

- Conference: International Conference on Stem Cell Engineering
- Year: 2016
- Proceeding: International Conference on Stem Cell Engineering 2016
- Group: General Submissions
- Session:

  Clinical Applications and Challenges (Including Scale Up and Imaging)

- Time:

  Tuesday, October 25, 2016 - 2:30pm-2:45pm

Authors:
Bulte, J. W., Johns Hopkins University

Abstract

Magnetic resonance (MR) imaging is expected to play a key role in evaluating the outcome of clinical trials based on stem cell therapy. In order to facilitate and implement the translation of these therapies into the clinic, it will be necessary to monitor the immediate cellular engraftment, subsequent biodistribution and migration, and cell survival and differentiation non-invasively over time. This information is simply not obtainable by invasive biopsy procedures that not only just provide a limited histological â□□snapshotâ□□, but may also be harmful for the patient.

MRI cell tracking, with its superior spatial resolution and excellent soft tissue anatomical detail, is now emerging as the technique of choice to monitor in real-time image-guided cell delivery and engraftment. Up until now, 10 clinical MRI cell tracking studies have been published, either using superparamagnetic iron oxide nanoparticles (SPIO) for proton (^{1}H) or perfluorocarbons for fluorine (^{19}F) MRI. SPIOs create strong local magnetic field disturbances that spoil the MR signal leading to hypointense contrast, while the fluorinated compounds generate signal â□□hot spotsâ□□ similar as those seen in nuclear medicine studies.

***** Formatting Issue - Best Available Paper/Graphic *****

Twelve years ago, our Cellular Imaging Section in the Johns Hopkins Institute for Cell Engineering was part of the team that performed the first-in-man ^{1}H MRI SPIO-labeled cell tracking study. Previous imaging studies used ^{111}In-oxine (radio)labeled cells and ultra-sound guided local tissue injection. A major surprise, and only revealed by MRI, was that the cells missed their target in half the patients. Our second clinical ^{1}H MRI SPIO-labeled stem cell tracking study in patients with amyotrophic lateral sclerosis (ALS) and multiple sclerosis (MS) showed that systemically and intrathecally injected mesenchymal stem cells (MSCs) homed into neuroinflammatory lesions in the brain and spinal cord, where they act as immunomodulators.

Together with Q Therapeutics Inc. and Dr. Nicholas Maragakis in Neurology we are now collecting pre-IND data for performing a clinical ^{19}F MRI tracking study of glial-restricted precursor cells in ALS. Lastly, funded by the Maryland Stem Cell Research Fund under the direction by Dan Gincel, we just obtained IND approval for a first-in man trial on ^{19}F MRI of perfluorocarbon-labeled stromal vascular fraction (SVF) cells for treatment of radiation-induced fibrosis in women with breast cancer. Together with CelSense Inc. and Dr. Ricardo Rodriguez at CosmeticSurg LLC, we will be scanning the first patients by the end of this year (NCT02035085).

Hyaluronic Acid Hydrogels Induce mTOR Activation to Enhance Isolated Rod Photoreceptor Survival in Vitro

5th International Conference on Stem Cell Engineering

- Conference: International Conference on Stem Cell Engineering
- Year: 2016
- Proceeding: International Conference on Stem Cell Engineering 2016
- Group: General Submissions
- Session:

  Towards Regenerative Therapies

- Time:

  Wednesday, October 26, 2016 - 12:00am-12:15am

Authors:
Mitrousis, N.
Tam, R. Y., University of Toronto
Baker, A. E. G., University of Toronto
van der Kooy, D., University of Toronto
Shoichet, M. S., University of Toronto

Research into photoreceptor development and degeneration has been greatly hindered by the lack of methods for *in vitro* culture of photoreceptors. Conventional 2-dimensional (2D) culture of photoreceptors leads to profound cell death and morphological changes. We found that culturing dissociated photoreceptors in hyaluronic acid (HA) â□□ based hydrogels greatly enhances their survival *in vitro*. We then investigated the molecular mechanism that mediates the prosurvival effect of the HA hydrogels on the photoreceptors.

Â

Rod photoreceptors were isolated by cell sorting from NrlGFP mice and cultured in 2D or in HA gels. Alive cells stained for Hoechst (nuclei), expressed GFP (Nrl-driven) and did not stain for EthD-1 (dead cell dye). To assess whether the differential stiffness of the HA gels compared to 2D tissue culture plastic mediated the prosurvival effect, we cultured photoreceptors in HA gels of varied stiffness. We found that altering substrate stiffness did not affect rod photoreceptor survival. To investigate the role of HA, we cultured photoreceptors in hydrogels without HA, or added soluble HA in 2D culture conditions. The presence of HA significantly enhanced rod survival, irrespective of whether the culture is performed in 2D or 3D. In order to decipher the molecular pathways induced by HA, immunostaining for phospho-mTOR as well as small inhibitor experiments were conducted. We used rapamycin, IWR1e and Y27632 to inhibit the mTOR, canonical Wnt and ROCK pathways, respectively. HA induced an increase in phospho-mTOR staining from 1.53% to 27.2%, and treatment with rapamycin abolished the prosurvival effect of the HA hydrogels. Treatment with IWR1e or Y26732 attenuated but did not eliminate the prosurvival effect of HA on rods. When either of IWR1e or Y26732 was combined with rapamycin, no additional decrease in rod survival was observed as compared to rapamycin alone. Lastly, when IWR1e and Y26732 were combined in the absence of rapamycin, the prosurvival effect of HA was abolished.

We conclude that the HA component of the HA-based hydrogels mediates photoreceptor survival through activating the mTOR pathway. The upstream mediators are canonical Wnt signaling and the RhoA/ROCK pathway. Our hydrogel represents a promising substrate for *in vitro* culture of stem cell-derived photoreceptors to study their development or perform screening assays. The pathways we identified may constitute compelling targets for investigation in animal models of retinal degeneration.

Regenerative Engineering of Human Tissues

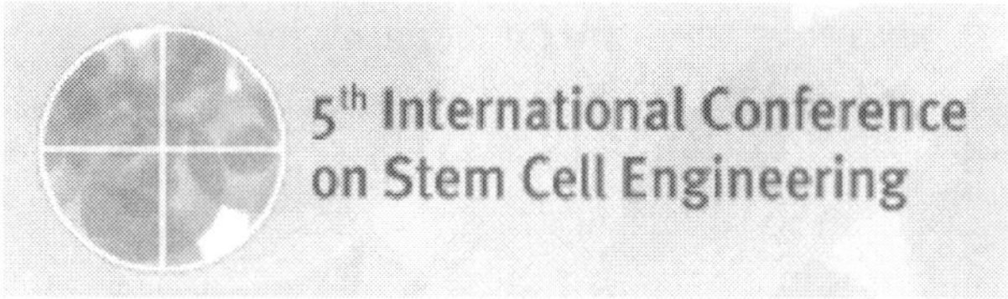

- Conference: International Conference on Stem Cell Engineering
- Year: 2016
- Proceeding: International Conference on Stem Cell Engineering 2016
- Group: General Submissions
- Session:

  Towards Regenerative Therapies

- Time:

  Wednesday, October 26, 2016 - 10:00am-10:30am

Authors:
Vunjak-Novakovic, G., Columbia University

Tissue engineering is becoming increasingly successful with authentically representing the actual environmental milieu of the development, regeneration and disease. A classical paradigm of tissue engineering is related to the integrated use of human stem cells, biomaterial scaffolds (structural and logistic templates for tissue formation) and bioreactors (culture systems providing environmental control, molecular and physical signaling) in regenerative medicine. Living human tissues can be bioengineered from the autologous stem cells, and tailored to the patient and the medical condition being treated. This talk will discuss some recent advances in regenerative engineering of bone, lung and heart.

*** Formatting Issue - Best Available Paper/Graphic ***

Scaling up a Chemically-Defined Aggregate-Based Suspension Culture System for Neural Commitment of Human Pluripotent Stem Cells

5th International Conference on Stem Cell Engineering

- Conference: International Conference on Stem Cell Engineering
- Year: 2016
- Proceeding: International Conference on Stem Cell Engineering 2016
- Group: General Submissions
- Session:

  Towards Regenerative Therapies

- Time:

  Wednesday, October 26, 2016 - 10:30am-10:45am

Authors:
Diogo, M. M., Department of Bioengineering and IBB - Institute for Biotechnology and Bioengineering, Instituto Superior Técnico, Universidade de Lisboa, Lisboa, Portugal
Miranda, C. C., Instituto Superior Técnico, Universidade de Lisboa
Fernandes, T. G., Instituto Superior Técnico, Universidade de Lisboa
Cabral, J. M. S., Instituto Superior Técnico, Universidade de Lisboa

The demand of high cell numbers for applications in cellular therapies and drug screening requires the development of scalable platforms capable to generating highly pure populations of tissue-specific cells from human pluripotent stem cells. In this work, we describe the scaling-up of an aggregate-based culture system for neural induction of human induced pluripotent stem cells (hiPSCs) under chemically-defined conditions. A combination of non-enzymatic dissociation and rotary agitation was successfully used to produce homogeneous populations of hiPSC aggregates with an optimal (140 μm) and narrow distribution of diameters (coefficient of variation of 21.6%). Scalable neural commitment of hiPSCs as 3D aggregates was performed in 50 mL spinner flasks, and the process was optimized using a factorial design approach, involving parameters such as agitation rate and seeding density. We were able to produce neural progenitor cell cultures, that at the end of a 6-day neural induction process contained less than 3% of Oct4-positive cells and that, after replating, retained more than 60% of Pax6-positive neural cells. The results here presented should set the stage for the future generation of a clinically relevant number of human neural progenitors for transplantation and other biomedical applications using controlled, automated and reproducible large-scale bioreactor culture systems.

3D Biomaterials for Accelerated Generation and Transplantation of Human Pluripotent Stem Cell Derived Midbrain Dopaminergic Neurons to Treat Parkinson's Disease

5th International Conference on Stem Cell Engineering

- Conference: International Conference on Stem Cell Engineering
- Year: 2016
- Proceeding: International Conference on Stem Cell Engineering 2016
- Group: General Submissions
- Session:

  Towards Regenerative Therapies

- Time:

  Wednesday, October 26, 2016 - 10:45am-11:00am

Authors:
Adil, M. M., University of California Berkeley
Rodrigues, G. M. C., Instituto Superior Técnico, University of Lisbon
Kulkarni, R. U., University of California Berkeley
Miller, E. W., University of California Berkeley

3D biomaterials for accelerated generation and transplantation of human pluripotent stem cell derived midbrain dopaminergic neurons to treat Parkinsons disease

Maroof M. Adil[1], Gonalo MC Rodrigues[2], Rishikesh U. Kulkarni, Tandis Vazin, Badriprasad Ananthanarayanan, Evan W. Miller, Sanjay Kumar and David V. Schaffer[1-4]

[1]Department of Chemical and Biomolecular Engineering; [2]Department of Bioengineering; [3]Department of Molecular and Cell Biology; [4]Helen Wills Neuroscience Institute; University of California, Berkeley, CA

Cell replacement therapies have considerable biomedical potential; however, low survival of cells post-transplantation is a major hurdle hampering their efficacy. For example, cell replacement therapies have strong promise for the treatment of neurodegenerative diseases such as Parkinsons, which is a currently incurable, progressively debilitating disorder affecting ~10M people in the world. However, following striatal transplantation of midbrain dopaminergic (mDA) neurons for the treatment of Parkinsons disease (PD), less than 5% of the neurons typically survive pre-clinically. Such poor neuronal survival requires the resource-intensive generation and implantation of larger numbers of mDA neurons to compensate, and the widespread death of the majority of the graft could reduce the efficacy of the cells remaining. We took a two-pronged approach to resolve this issue by developing methods to first increase the yield of human pluripotent stem cell (hPSC) derived mDA neurons generated in vitro, and second, to increase the post-transplantation survival of neurons in vivo.

Human pluripotent stem cells (hPSCs) were cultured and differentiated into mDA neurons in 3D pNIPAAm-PEG gels. The authenticity of resulting neurons thoroughly characterized by immunocytochemistry, qPCR and voltage-sensitive dye based electrophysiology, and benchmarked against mDA neuronal development on 2D surfaces. Subsequently, a biodegradable, hyaluronic acid (HA) based hydrogel functionalized with ECM derived factors was developed for maturation and transplantation of mDA neurons generated on pNIPAAm-PEG. Post-transplantation survival of hydrogel-encapsulated neurons was then investigated 4.5 months after striatal implantation into rats.

mDA neuron production typically involves undefined components and difficult to scale-up 2D culture formats. Here, we used a fully-defined, 3D thermoresponsive biomaterial platform to derive mDA neurons from PSCs at substantially greater yield, quality, and maturity compared to 2D Matrigel coated surfaces. In particular, after 25 days of differentiation, ~40% of cells differentiated in 3D expressed tyrosine hydroxylase (TH), the rate limiting enzyme for dopamine production and a hallmark of mDA neurons, compared to 20% on 2D. Crucial for functional mDA neurons, ~ 5-fold higher proportion of neurons differentiated in 3D fired mDA-like action potentials. Interestingly, FOXA2 and EN1, markers known to improve survival of mDA neurons, were expressed at significantly higher levels in cells generated in the 3D platform compared to control 2D surfaces.

Low post-transplantation cell survival is often attributed to adverse biochemical, mechanical, and immunological stresses cells experience during and post-transplantation. To address these challenges, we developed a functionalized hyaluronic acid (HA) based hydrogel first to mature pluripotent stem cell (PSC) derived neural progenitors towards functional neurons in vitro in a biomimetic 3D environment, and second to effectively transplant hydrogel-encapsulated neurons into the central nervous system. We demonstrated that incorporating adhesive peptides to promote cell adhesion increased extended neurite density, and adding neurotrophic factors increased dopaminergic differentiation in our 3D HA gels. Furthermore, we observed higher survival post-harvest and injection in vitro for mDA neurons matured within hydrogels compared to those on 2D surfaces. By using an optimized 3D HA biomaterial matrix for neuronal maturation and transplantation, we significantly increased the post-transplantation survival of hPSC derived mDA neurons in the rat striatum by 5 fold compared to the current standard method of bolus injections of unencapsulated neurons.

The combination of a defined, scalable, and resource-efficient cell culture platform and an optimally engineered transplantation biomaterial significantly increased the yield of mDA neurons generated in vitro and improved their post-transplantation viability in vivo, thereby addressing crucial challenges in treatment of neurodegenerative disease.

Subcutaneously Transplanted Islet-Embedded Endothelialized Modules Restore Normoglycemia in Diabetic SCID/Bg Mice

5th International Conference on Stem Cell Engineering

- Conference: International Conference on Stem Cell Engineering
- Year: 2016
- Proceeding: International Conference on Stem Cell Engineering 2016
- Group: General Submissions
- Session:

  Towards Regenerative Therapies

- Time:

  Wednesday, October 26, 2016 - 11:00am-11:15am

Authors:
Vlahos, A. E., University of Toronto
Sefton, M. V., University of Toronto
Cober, N.

Subcutaneously transplanted islet-embedded endothelialized modules restore normoglycemia in diabetic SCID/Bg mice

The transplantation of pancreatic islets via the hepatic portal vein (Edmonton Protocol)1 is a promising treatment for type 1 diabetics1. However, this therapy has limitations2, two of which include: (1) the need for multiple donors for each transplant to achieve insulin independence3 and (2) insulin independence is not sustained2. Finding an alternative transplant site that is both minimally invasive and able to support a large transplant volume is necessary for restoring normoglycemia in diabetes. Although the subcutaneous site satisfies both of these criteria, the site is poorly vascularized. Modular tissue
engineering is a â□□bottom-upâ□□ approach which uses collagen gel rods, mesenchymal stromal cells (MSC) and endothelial cells (EC) to form a subcutaneous vascularized bed and can be used to deliver pancreatic islets4. In this study, the

subcutaneous transplantation of Wistar rat islets (750 IEQ, islet equivalents) embedded in endothelialized modules (mixed with EC modules, with or without MSC) were sufficient to restore and maintain normoglycemia in streptozotocin induced diabetic SCID/Bg mice for up to 21 days. Free islets had no effect. Islets embedded in collagen but without endothelial cells were able to return diabetic animals to normoglycemia, but insulin independence was not sustained (animals became hyperglycemic within 21 days). Islets embedded within endothelialized modules were able to drive vascularization of the subcutaneous implant by recruiting host-derived vessels within 14 days of transplantation.

In addition, we assessed the intra-islet vasculature and found that embedded islets in endothelialized modules had become re-vascularized and integrated with the hostâ□□s vasculature. The presence of an integrated and perfuseable vasculature is not a feature of previous the relatively few previous subcutaneous implant studies(e.g., 5).

Embedded islets drove a more robust and prolonged inflammatory response observed by a large recruitment of neutrophils at day 7. A shift towards a CD206+MHCII- (M2-like) macrophage was also observed at day 7 in islet-embedded modular implants, which continued to day 14. While we have demonstrated that module-associated vasculature is sufficient to support islet function and viability over 21 days, it remains to reduce the number of islets required to achieve normoglycemia. To this end, we are using modules to pre-vascularize the subcutaneous site before islet transplantation.

References:

1. Shapiro, A. M. J. *et al.* Islet Transplantation in Seven Patients with Type 1 Diabetes Mellitus Using a Glucocorticoid-Free Immunosuppressive Regimen. *N. Engl. J. Med*

343, 230â□□238 (2000).

2. McCall, M. & James Shapiro, A. M. Update on Islet Transplantation. *Cold Spring*

Harbor Perspectives in Medicine **2,** a007823â□□a007823 (2012).

3. Markmann, J. F., Deng, S., Huang, X. & Desai, N. M. Insulin independence following isolated islet transplantation and single islet infusions. *Annals of â□¦* **237,** 741â□□750 (2003).

4. Butler, M. J. & Sefton, M. V. Cotransplantation of Adipose-Derived Mesenchymal Stromal Cells and Endothelial Cells in a Modular Construct Drives Vascularization in SCID/bg Mice. *Tissue Engineering Part A* **18,** 1628â□□1641 (2012).

5. Pepper, A. R. *et al.* A prevascularized subcutaneous device-less site for islet and cellular transplantation. *Nat Biotechnol* **33,** 518â□□523 (2015).

Rapid Vascularization of Engineered Tissues

5th International Conference on Stem Cell Engineering

- Conference: International Conference on Stem Cell Engineering
- Year: 2016
- Proceeding: International Conference on Stem Cell Engineering 2016
- Group: General Submissions
- Session:

  Towards Regenerative Therapies

- Time:

  Wednesday, October 26, 2016 - 11:15am-11:30am

Authors:
Grayson, W., Johns Hopkins University

The translation of tissue engineering strategies from bench-side to clinical therapeutics is significantly hampered by the inability to provide adequate oxygen to support the post-transplantation metabolic needs of the tissue. To overcome this limitation, our lab is developing approaches to enable rapid vascularization of the engineered orthopaedic and cardiovascular tissues. We incorporate heterogenous populations of human adipose-derived stem/stromal cells (hASCs), which have demonstrated potential to spontaneously self-assemble into robust, interconnected vascular networks and/or stabilize neo-vessels formed by endothelial cells. Used in combination with differentiated cell phenotypes and advanced functionalized biomaterials, we will demonstrate various roles hASCs may adopt to facilitate the growth of complex vascularized tissues - including bone, skeletal muscle, and myocardium - without inhibiting cell and tissue function.

Endogenous Cell Stimulation and Exogenous Cell Transplantation to Promote Repair in the Central Nervous System

5th International Conference on Stem Cell Engineering

- Conference: International Conference on Stem Cell Engineering
- Year: 2016
- Proceeding: International Conference on Stem Cell Engineering 2016
- Group: General Submissions
- Session:

  Plenary Session

- Time:

  Wednesday, October 26, 2016 - 12:00am-12:25am

Authors:
Shoichet, M. S., University of Toronto
Tuladhar, A., University of Toronto
Samantha, P., University of Toronto
Nagy, A.
Morshead, C., University of Toronto

Regenerative strategies involving endogenous stem cell stimulation or exogenous stem cell transplantation provide promise in devastating diseases and traumatic injuries of the central nervous system, such as stroke. With only tissue plasminogen activator approved for the treatment of stroke, there is significant opportunity to design alternative strategies; however, the blood-brain barrier poses a significant barrier, limiting access to the brain using traditional intravenous and oral delivery strategies.

In order to overcome the BBB, we have designed a local delivery strategy that circumvents this barrier. By delivering therapeutic biomolecules directly to the brain tissue, we propose to stimulate the endogenous stem cells that line the lateral ventricles. We build on previous research where the sequential, co-delivery of epidermal growth factor (EGF) and erythropoietin (EPO) successfully promoted tissue repair in an endothelin-1 mouse model of stroke [1]. Here, we test endogenous stem cell stimulation in a larger, endothelin-1 rat model of stroke with the local delivery of two molecules that promote neural precursor activation: cyclosporine A and EPO. The two molecules are separately encapsulated in PLGA microspheres and dispersed in a hyaluronan/methylcellulose (HAMC) hydrogel, which is then injected directly on the cortex. We examine tissue and locomotor functional repair over 6 weeks.

Simultaneously, in a separate set of studies, we investigate cell transplantation strategies. We build on previous research where mouse neural precursor cells injected into the lesion site promoted tissue and functional repair [2]. Cells injected in the HAMC hydrogel showed greater cell survival and functional repair relative to controls. Here, we test transplantation of human-induced pluripotent stem cells differentiated to cortically-specified neuroectodermal precursor cells. We investigate the role of *in vitro*neuronal maturation on transplanted cell survival, tissue and functional repair in a rat endothelin-1 model of stroke.

In these studies of endogenous stem cell stimulation and exogenous stem cell transplantation to the central nervous system, we tackle key challenges of the field including cell survival and cell integration. Our strategies at the intersection of engineering and biology will be highlighted.

Acknowledgments: We are grateful to: the Canadian Institute for Health Research (CIHR) and Medicine by Design for funding (to MSS, CM and AN), the Natural Sciences and Engineering Research Council for a graduate student scholarship (to SP), and to the Ontario Graduate Scholarship and CIHR TPRM for graduate student scholarships (to AT)

References:

[1] Wang, Y.; Cooke, M.J.; Sachewsky, N; Morshead, C.M.; Shoichet, M.S. 2013 "Bioengineered sequential growth factor delivery stimulates brain tissue regeneration after stroke", J Control Rel, 172: 1-11

[2] Ballios, B.G.; Cooke, M.J.; Clarke, L.; Cole, B.; Morshead, C.M.; van der Kooy, D.; Shoichet, M.S. 2015 "A hyaluronan based injectable hydrogel improves the survival and integration of stem cell progeny following transplantation" Stem Cell Reports, 4: 1031-45

Biomaterial mechanics, stem cells and regeneration

5th International Conference on Stem Cell Engineering

- Conference: International Conference on Stem Cell Engineering
- Year: 2016
- Proceeding: International Conference on Stem Cell Engineering 2016
- Group: General Submissions
- Session:

  Plenary Session

- Time:

  Wednesday, October 26, 2016 - 12:25am-12:50am

Authors:
Mooney, D. J., Harvard University

Physical cues can control stem cell behavior, but this has not been widely exploited to date in regeneration. We have now found that stem cell fate and ability to promote regeneration is highly regulated by both matrix stiffness and stress relaxation. In parallel, materials that provide mechanical stimulation, on-demand, directly to damaged tissue have been found to promote skeletal muscle regeneration even after severe injury. The ability of biomaterials, via their intrinsic mechanical properties and mechanical stimulation of injured tissues, to promote regeneration suggests new paradigms for regenerative medicine.

*** Formatting Issue - Best Available Paper/Graphic ***

3D Cerebellar Differentiation and Purkinje Cell Maturation from Human Induced Pluripotent Stem Cell

5th International Conference on Stem Cell Engineering

- Conference: International Conference on Stem Cell Engineering
- Year: 2016
- Proceeding: International Conference on Stem Cell Engineering 2016
- Group: Poster Submissions
- Session:

  Poster session

- Time:

  Tuesday, October 25, 2016 - 5:30pm-7:30pm

Authors:
Fernandes, T. G., Instituto Superior Técnico, Universidade de Lisboa
Diogo, M. M., Department of Bioengineering and IBB - Institute for Biotechnology and Bioengineering, Instituto Superior Técnico, Universidade de Lisboa, Lisboa, Portugal
Rodrigues, C. A. V., Instituto Superior Técnico, University of Lisbon
Cabral, J. M. S., Instituto Superior Técnico, Universidade de Lisboa
Silva, T. P., Department of Bioengineering and iBB – Institute for Bioengineering and Biosciences, Instituto Superior Técnico, Universidade de Lisboa
Bekman, E.
Carmo-Fonseca, M.

The cerebellum plays a critical role in maintenance of balance and posture, coordination of voluntary movements and motor learning. Cerebellar ataxias represent a heterogeneous group of disorders characterized by progressive neurodegeneration of the cerebellum for which no cure is currently available. Most of the current knowledge about ataxias is based on postmortem studies and animal models that do not fully recapitulate the features of human disease. The development of new models to study cerebellar ataxia is therefore an important unmet medical need. In recent years, discoveries regarding the mechanism of cerebellar differentiation has promoted the generation of cerebellar neurons from pluripotent stem cells[1]. However, despite recent progress, the establishment of long-term culture systems and generation of different mature cerebellar neurons remains a challenge.

In the present study, we recapitulate the early developmental events of human cerebellum in vitro from human iPSCs in 3D, with generation of continuous cerebellar plate neuroepithelium. For this propose, cells were differentiated into cerebellar neurons using a 3D culture system and sequential addition of growth factors, including FGF2, FGF19 and SDF1[1]. FGF2 has an inductive role in cerebellar differentiation, with a major effect in the caudalization of neuroepithelium tissue. FGF19 was necessary for promoting spontaneous generation of rostral hindbrain neural-tube-like structures with dorsal-ventral (D-V) polarity, and SDF1 was shown to facilitate spontaneous generation of rhombic lip-like structure as seen at the developmental stage when cerebellar neurogenesis occurs. Our system generated a neuroepithelium organized into a multilayered structure with basal-apical polarity, including Rhombic-lip-derivative zone (BARHL1+), Purkinje cells precursors zone (Olig2+ and SKOR2+) and Ventricular zone (Sox2+ and N-cad+), characteristic to the cerebellar ontogenesis. Furthermore, we are able to maintain this 3D culture until 134 days of differentiation and obtain MAP2^{+} and calbindin^{+} mature neurons.

Given the importance of Purkinje cells in cerebellar neuropathology, we proceeded to isolate these cells without the need for co-culture with other cell types or cerebellum slices, as has been previously demonstrated[1,2]. To this end we dissociated 3D organoids and replated the cells on laminin in a defined serum-free neuronal basal medium reported to be more representative of the central nervous system extracellular environment and increasing cell survival and the proportion of synaptically active neurons[3]. Replated cells show typical markers of Purkinje cells (calbindin), and were able to organize into intricate neuronal networks. Cells were maintained for 134 days in culture and show signs of maturation.

To our best knowledge, this is the first report that shows the culture of Purkinje cells without co-culture and using defined conditions. Thus, we have established an important foundation to generate cultures of cerebellar neurons that can be used for further studies on cerebellar ataxias.

1. Muguruma, K., Nishiyama, A., Kawakami, H., Hashimoto, K. & Sasai, Y. Self-Organization of Polarized Cerebellar Tissue in 3D Culture of Human Pluripotent Stem Cells. *Cell Rep.* **10,** 537â□□550 (2015).

2. Wang, S. *et al.* Differentiation of human induced pluripotent stem cells to mature functional Purkinje neurons. *Sci. Rep.* **5,** 9232 (2015).

3. Bardy, C. *et al.* Neuronal medium that supports basic synaptic functions and activity of human neurons in vitro. *Proc. Natl. Acad. Sci. U. S. A.* **112,** E2725â□□34 (2015).

A Novel Noninvasive, Quantitative Molecular Imaging Platform for Assessing Transcriptional States during Endoderm and Hepatocyte Formation from Human Pluripotent Stem Cells

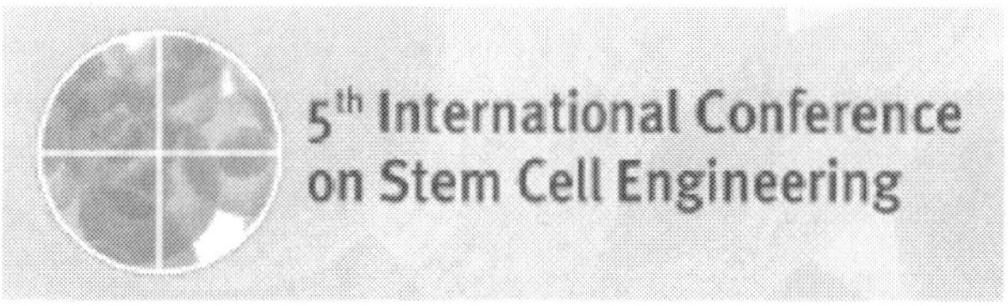

- Conference: International Conference on Stem Cell Engineering
- Year: 2016
- Proceeding: International Conference on Stem Cell Engineering 2016
- Group: Poster Submissions
- Session:

  Poster session

- Time:

  Tuesday, October 25, 2016 - 5:30pm-7:30pm

Authors:
Willadsen, M.
Meamardoost, S., University at Buffalo
Chase, M.
Paulmurugan, R.
Fisher, S.
Parashurama, N., University of California, San Francisco

Although the liver regenerates in vivo, the regenerative function is lost during fibrosis, and hepatocytes donâ□□t divide in culture. Pluripotent stem cells (PSCs) are an exciting potential source of hepatocytes, but thus far scientists have failed to generate fully mature hepatocytes. Hepatocytes are derived from endoderm, one of three main germ layers during development. Primary human endoderm, and thus human pluripotent stem cell (PSC)-derived endoderm, remain an enigma, and PSC endoderm studies demonstrate phenotypic heterogeneity, aberrant differentiation, difficulty with maturation, and in some cases, resistance to differentiation. In vivo genetic studies has demonstrated that developmental, master, regulatory gene circuits (DRGC), composed of transcription factors, govern endoderm differentiation. We hypothesized that a poorly regulated DRGC may be root cause for the problems with liver differentiation. To better understand DRGC, we developed a noninvasive, quantitative platform for imaging endoderm in vitro and in vivo. We engineered endoderm transcription factor (TF) -specific response elements which contain DNA binding sites for Brachyury, Foxa2, and Sox17, three key TFs that compose the DRGC in endoderm. The binding sites were flanked upstream by sequences a two-step transcriptional activation system, to improve signal strength in baseline and stimulated states. Our data indicated that low levels of signal were stable at 6 weeks after lentiviral transduction, enabling easy colony identification. We initiated PSC differentiation for three days in coated matrigel plates, and Activin, a key soluble factor that induces the DRGC as a function of concentration. We imaged at least 3 and up to 15 colonies a condition, and concentration experiments (0, 2, 20, 40, 100 ng/ml) were performed in duplicate. We normalized reporter expression to control hPSC which ubiquitously express luciferase. In a total of 24 PSC colonies containing the reporters, max radiance was $2 \times 10^5 \pm 1.6\ 10^5$ photons/cm^2-s/sr, demonstrating considerable baseline signal strength. The Brachyury reporter activity demonstrated decreasing activity with increasing concentration, and 42% decrease in normalized signal at 100 ng/ml compared to 0 ng/ml. The Foxa2 reporter demonstrated a 143% increase in normalized signal at 100 ng/ml compared to 0ng/ml. Finally, the Sox17 reporter demonstrated a 91% increase in normalized signal at 100 ng/ml compared to 0 ng/ml. These data are consistent with expected result, which is a downregulation of Brachyury, the precursor to mesendoderm, and upregulation of Foxa2 and Sox17, which are together specific for endoderm, in the presence of Activin. Data for intermediate concentrations of Activin were inconclusive. Future work will focus on correlating reporter expression with RNA or protein levels, repeating experiments at intermediate concentrations of Activin, improving clonality of cell lines, and studying transcriptional events in vivo. These studies will lead to a greatly improved understanding of the DRGC which govern endoderm induction and differentiation along the hepatic lineage for generating functional hepatocytes.

Angiogenic Secretome of Xeno-Free Hasc

5th International Conference on Stem Cell Engineering

- Conference: International Conference on Stem Cell Engineering
- Year: 2016
- Proceeding: International Conference on Stem Cell Engineering 2016
- Group: Poster Submissions
- Session:

  Poster session

- Time:

  Tuesday, October 25, 2016 - 5:30pm-7:30pm

Authors:
Escobar, C.
Chaparro, O.
Background/aims. Molecules of animal or bacterial origin, which pose a risk of zoonoses and/or immune rejection, are commonly used for extraction, culture and cryopreservation of mesenchymal stem cells (MSCs). To evaluate qualitatively, quantitatively and functionality, the xeno-free hASC neovascularization secretome, is a good way to evaluate the revascularization capacity of these cells, and these results could suggest the effect of the xeno-free production and the cell culture time on their regenerative potential.Â ***Methods.*** The neovascularization secretome of hASC from two samples in fourth and seventh passage, produced under xeno-free and standard conditions were evaluated using an antibody array andÂ in vitro, ex vivo and in vivo, angiogenesis functional assays. ***Results*** The xeno-free hASC secrete higher concentrations of PDGF-BB and IFN-Î³. The fourth passage xeno-free hASC secrete higher concentration of several proangiogenic factors like CCL-13, MMP-1 and 9, VEGF and TGF- Î²1. The seventh passage xeno-free hASC secrete higher concentration of CXCL-5, VEGFA and D, THPO and VEGFR3. The seventh passage xeno-free hASC, shows an increased capacity in the in vitro and ex vivo angiogenesis assays, and the fourth passage shows also increase in the ex vivo angiogenesis assay. In the hind limb ischemia assay there are no substantial differences between the standard and the xeno-free produced cells. ***Conclusions.*** The xeno-free production changed quantitatively the hASC secretome and causes an increased neovascularization capacity of it. Under standard conditions, the cell culture time induces an increase in the secretion of antiangiogenic factors, which is correlated with a reduced neovascularization capacity.

Biodegradable Scaffold with Built-in Vasculature for Organ-on-a-Chip Engineering and Direct Surgical Anastomosis

5th International Conference on Stem Cell Engineering

- Conference: International Conference on Stem Cell Engineering
- Year: 2016
- Proceeding: International Conference on Stem Cell Engineering 2016
- Group: Poster Submissions
- Session:

  Poster session

- Time:

  Tuesday, October 25, 2016 - 5:30pm-7:30pm

Authors:
Zhang, B., University of Toronto
Radisic, M., University of Toronto
Montgomery, M.
Chamberlain, D.
Ogawa, S.
Korolj, A., University of Toronto
Panhke, A., University of Toronto
Wells, L.
Massé, S.
Kim, J.
Reis, L.
Momen, A.
Nunes, S.
Wheeler, A.
Nanthakumar, K., Toronto General Research Institute
Keller, G., McEwan Centre for Regenerative Medicine
Sefton, M. V., University of Toronto

Biodegradable scaffold with built-in vasculature for organ-on-a-chip engineering and direct surgical anastomosis

---

1,2Boyang Zhang, 1,2Miles Montgomery, 2M. Dean Chamberlain, 7Shinichiro Ogawa, 1,2Anastasia Korolj, 1,2Aric

Pahnke, 2Laura A. Wells, 5StÃ©phane MassÃ©, 3Jihye Kim, 2Lewis Reis, 4Abdulah Momen, 2,4,6Sara S. Nunes,

1,3Aaron R. Wheeler, 5Kumaraswamy Nanthakumar, 7Gordon Keller, 1,2Michael V. Sefton, and 1,2,4,6,*Milica

Radisic

1Department of Chemical Engineering and Applied Chemistry, University of Toronto, Toronto, Ontario, Canada.

2Institute of Biomaterials and Biomedical Engineering, University of Toronto, Toronto, Ontario, Canada.

3Department of Chemistry, University of Toronto, Toronto, Ontario, Canada. 4Toronto General Research Institute, University Health Network, Toronto, Ontario, Canada. 5The Toby Hull Cardiac Fibrillation Management

Laboratory, Toronto General Hospital, Toronto, Ontario, Canada.6The Heart and Stroke/Richard Lewar Centre of

Excellence, Toronto, Ontario, Canada. 7McEwen Center for Regenerative Medicine, Toronto, Ontario, Canada

Recapitulating vascular interfaces of different organs in 3-D is critical in both organ-on-a-chip and tissue engineering applications. Vascular networks can be engineered with subtractive fabrication by embedding a sacrificial carbohydrate-glass lattice, Pluronic F127, or dry alginate fibers in hydrogels. However, the soft hydrogel provides only a temporary structural support for the fragile hollow network and does not permit extensive tissue remodeling, which inevitably alters the hydrogel structure and collapses the embedded network. Synthetic biodegradable polymers could provide sufficient structural support to the engineered vessels, but their low permeability prevents biomolecule exchange and cell migration between the vessels and the parenchymal space.

To accommodate these two opposing material criteria we created AngioChip, constructed using a biodegradable elastomer, poly(octamethylene maleate (anhydride) citrate) (POMaC). POMaC was selected since it is UV- polymerizable, allowing rapid assembly under mild conditions, biodegrades by hydrolysis, and is more elastic than FDA approved

polyesters. The stable biodegradable scaffold includes a built-in branching micro-channel network that contained two unique features realized by our new 3-D stamping technique. First, the synthetic
built-in vascular walls were thin and flexible, yet strong enough to mechanically support a perfusable vasculature
in a contracting tissue and enable direct surgical anastomosis. Second, to allow efficient molecular exchange and cell migration, nano-pores and micro-holes were incorporated into the vascular walls. By establishing a stable, permeable, vessel network within AngioChips, we were liberated from material constraints, which allowed us to use any soft natural extracellular matrix (e.g. collagen, Matrigel) embedded with cells in the parenchymal space permitting the extensive tissue remodelling.
Cardiac tissues were created from human embryonic stem cell (hESC), human mesenchymal stem cells (MSCs) as a side population and HUVECs for inner lumen coating. To provide an evidence of tissue-level organization, confocal images and histological cross-sections show compact layers of cells throughout the entire tissue volume, including the scaffold interior. Fully human liver-AngioChips were engineered using hESC derived hepatocytes. High-density culture resulted in the formation of junctions between hepatocytes, and a positive staining for albumin and bile canaliculi. Vascularized hepatic tissues and cardiac tissues, engineered using AngioChips, were shown to process clinically relevant drugs delivered through their internal vasculature. Incorporation of nano-pores and micro-holes in the vessel walls enhanced vessel permeability, permitted inter-cellular crosstalk, extravasation of monocytes and sprouting of endothelial cells upon stimulation. AngioChip cardiac tissues were also implanted via direct surgical anastomoses to the femoral vessels of rat hindlimbs, establishing immediate blood perfusion.

*** Formatting Issue - Best Available Paper/Graphic ***

Bioengineered Hyaluronan-Based Hydrogels for the Delivery of Retinal Cell Precursors

5th International Conference on Stem Cell Engineering

- Conference: International Conference on Stem Cell Engineering
- Year: 2016
- Proceeding: International Conference on Stem Cell Engineering 2016
- Group: Poster Submissions
- Session:

  Poster session

- Time:

  Tuesday, October 25, 2016 - 5:30pm-7:30pm

Authors:
Delplace, V., University of Toronto
Shoichet, M. S., University of Toronto
van der Kooy, D., University of Toronto
Ballios, B. G., Chemical Engineering & Applied Chemistry
Wallace, V. A.
Bergeret, M.
Tam, R. Y., University of Toronto
Cooke, M. J., The University of Manchester

Bioengineered hyaluronan-based hydrogels for the delivery of retinal cell precursors

Introduction. Blindness and visual impairment are largely caused by irreversible loss of light sensitive photoreceptors in the retina. Transplantation of photoreceptor cells and their progeny is one of the most promising strategies to restore vision [1]. But, using conventional injection strategies, the distribution, survival and integration of such cells in vivo remains a major impediment to clinical applications. To enhance retinal cell distribution in vivo, we designed an injectable hyaluronan-based hydrogel that met with much success [2-7]. Yet, 90% (or more) of transplanted cells are still dying. In order to make a more significant difference, specific interactions between photoreceptor cells and their natural microenvironment were investigated as a source of inspiration for innovative biomaterials. Following new progress in the biological understanding, further chemical modifications of HAMC with the appropriate pro-survival factors are being investigated to deliver the cells in a more conducive environment for even greater survival and integration.

Materials and methods. A hydrogel formulation comprised of a physical blend of hyaluronan (HA) and methylcellulose (MC), HAMC, was developed and optimized for subretinal delivery. The injectable hydrogel was tested in vivo for transplantation of retinal cells and their photoreceptor progeny. Then, using a new model of GFP-labeled cone-like cells from Nrl-/- ccdc136GFP mice [8], cell surface receptors were characterized by immunostaining, and cell behavior in 2D on various interphotoreceptor matrix (IPM) components was investigated. HAMC was chemically modified to allow thiol-maleimide and biotinâ□□ streptavidin chemistry, and conjugated with various pro-survival factors. The modified hydrogel is being tested with photoreceptor cells in vitro and in vivo.

Results and discussion. The HAMC hydrogel is particularly well-suited for delivery into the subretinal space as it is minimally swelling, bioresorbable within one week and it gels rapidly on injection [5-7]. When delivered in HAMC, the photoreceptor progeny was well-distributed in the subretinal space after injection (see **Figure 1**), and significantly better than those injected using conventional saline strategies, where the cells aggregated together [2-4]. Cell survival was doubled from approximately 4% to 8% when delivered in HAMC vs. saline controls.

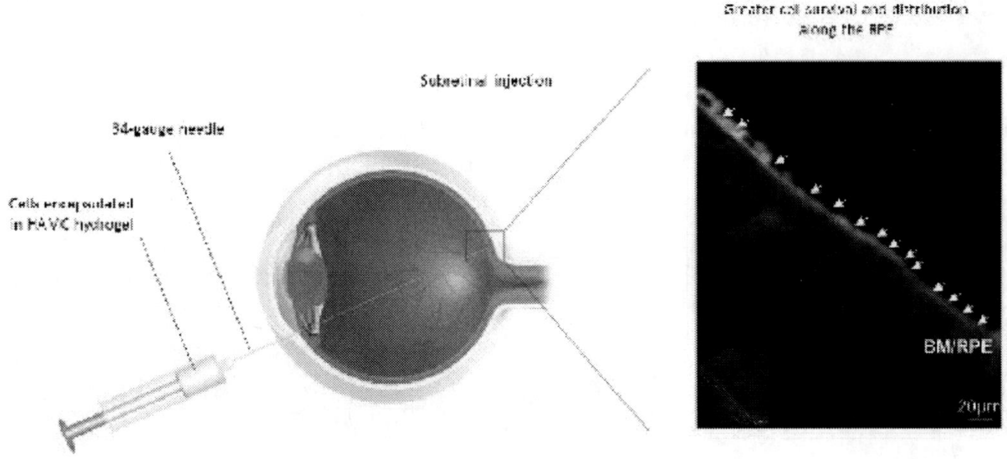

Figure 1. Delivery of retinal progenitor cells (RPCs) to the subretinal space, using hyaluronan/methylcellulose (HAMC) hydrogel as biomaterial support.

1

To further enhance the properties of HAMC, the hydrogel could be covalently modified with biomimicking pro-survival factors. However, few is known about the interactions between the photoreceptor cells and their microenvironment. Taking advantage of a most recently developed gene trap for the specific GFP-labeling of cones and Nrl-/- photoreceptor cells, the presence of laminin-binding integrins was assessed, and the specific bioactivity of laminin compared to other adhesive glycoproteins (i.e., fibronectin and vitronectin) on photoreceptors was confirmed in vitro, in 2D conditions. HAMC was then covalently modified with two laminin-inspired synthetic peptides. The success of the synthesis was assessed by mass spectroscopy and amino acid analysis, prior to a complete study of photoreceptor cell viability first in vitro and then in vivo. The HAMC hydrogel, conjugated with pro-survival factors, should enhance both distribution and survival of transplanted retinal photoreceptor cells.

Conclusions. These data demonstrate that HAMC serves as a platform strategy for cell delivery, enabling minimally-invasive strategies and significantly better tissue and functional outcomes than controls.

References

[1] V. Delplace, S. Payne, M. Shoichet, Delivery strategies for treatment of age-related ocular diseases: from a biological understanding to biomaterial solutions., J. Control. Rel., (2015).

[2] B.G. Ballios, M.J. Cooke, L. Donaldson, B.L. Coles, C.M. Morshead, D. van der Kooy, M.S. Shoichet, A Hyaluronan- Based Injectable Hydrogel Improves the Survival and Integration of Stem Cell Progeny following Transplantation, Stem Cell Reports, 4 (2015) 1031-1045.

[3] B.G. Ballios, M.J. Cooke, D. van der Kooy, M.S. Shoichet, A hydrogel-based stem cell delivery system to treat retinal degenerative diseases, Biomaterials, 31 (2010) 2555-2564.

[4] R.Y. Tam, T. Fuehrmann, N. Mitrousis, M.S. Shoichet, Regenerative therapies for central nervous system diseases: a biomaterials approach, Neuropsychopharmacology, 39 (2014) 169-188.

[5] H. Kim, M.J. Cooke, M.S. Shoichet, Creating permissive microenvironments for stem cell transplantation into the central nervous system, Trends Biotechnol., 30 (2012) 55-63.

[6] D. Gupta, C.H. Tator, M.S. Shoichet, Fast-gelling injectable blend of hyaluronan and methylcellulose for intrathecal, localized delivery to the injured spinal cord, Biomaterials, 27 (2006) 2370-2379.

[7] M.D. Baumann, C.E. Kang, C.H. Tator, M.S. Shoichet, Intrathecal delivery of a polymeric nanocomposite hydrogel after spinal cord injury, Biomaterials, 31 (2010) 7631-7639.

[8] S. Smiley, P.E. Nickerson, L. Comanita, N. Daftarian, A. El-Sehemy, E.L.S. Tsai, S. Matan-Lithwick, K. Yan, S. Thurig, Y. Touahri, R. Dixit, T. Aavani, Y. De Repentigny, A. Baker, C. Tsilfidis, J. Biernaskie, Y. SauvÃ©, C. Schuurmans, R. Kothary, A.J. Mears, V.A. Wallace, Establishment of a cone photoreceptor transplantation platform based on a novel cone-GFP reporter mouse line, Scientific Reports, 6 (2016) 22867.

*** Formatting Issue - Best Available Paper/Graphic ***

Biomimetic Self-Assembled Scaffolds Enhance Muscle Stem Cell Transplantation

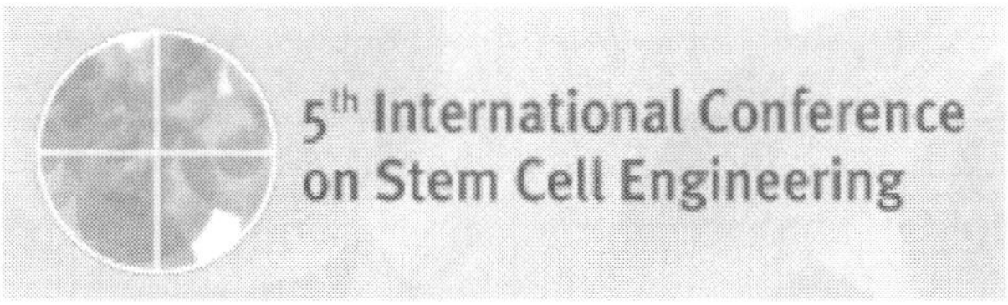

- Conference: International Conference on Stem Cell Engineering
- Year: 2016
- Proceeding: International Conference on Stem Cell Engineering 2016
- Group: Poster Submissions
- Session:

  Poster session

- Time:

  Tuesday, October 25, 2016 - 5:30pm-7:30pm

Authors:
Cosgrove, B. D., Cornell University
Blau, H. M., Stanford University
Stupp, S. I., Northwestern University
Meade, T. J., Northwestern University
Haynes, R. D., Stanford University School of Medicine
Sleep, E., Northwestern University
McClendon, M. T., Northwestern University
Preslar, A. T., Northwestern University

Muscle stem cells (also known as satellite cells) are essential for endogenous skeletal muscle regeneration but their utility in cell transplantation therapy is currently limited by their inefficient survival, self-renewal, and differentiation after injection into muscle tissue. To improve the contributions of MuSCs post-transplant, we developed biomimetic scaffolds based on synthetic peptide amphiphiles (PAs) that self-assemble to encapsulate cells and growth factors within a muscle-like unidirectionally-ordered environment of long and aligned nanofibers directly as they are injected into damaged muscle tissue.

We generated PAs consisting of an aliphatic (C_{16}) palmitoyl tail and hydrophilic nine amino acid cap (e.g. $V_3A_3E_3$, in various rearrangements) using a peptide synthesizer. We annealed PAs into randomly-ordered, long-axis nanofibers, which externally expose the amino acid cap and internally aggregate the hydrophobic tail, by entropic mixing. We extruded annealed PAs (aPAs) into physiological calcium concentrations in culture medium to fabricate a stable, noodle-like aPA scaffolds with highly-ordered nanofibers in parallel to the direction of extrusion. We measured the storage modulus of aPA scaffolds using shear rheometry and observed shear storage moduli (Gâ□□) spanning three orders of magnitude, from Gâ□□ = 3-20 kPa, encompassing a range of physiological muscle stiffness.

We evaluated myogenic progenitor cell viability, proliferation and differentiation after encapsulation in assembled aPA scaffolds *in vitro*. We observed that the aPA scaffold stiffness determined the macroscopic degree of muscle progenitor cell alignment with the ordered nanofiber template. The aPA scaffolds supported primary myoblast survival and proliferation *in vitro*, and, when tuning their stiffness to Gâ□□ = 9 kPa, optimal myoblast differentiation into mature myosin heavy chain-positive myofibers.

Upon injection into hindlimb muscles, we observed longitudinal co-alignment of scaffold nanofibers with recipient muscle myofibers to achieve desired templating of injected cells into differentiation tracks in parallel with the native architecture. aPA scaffold degradation *in vivo* was assayed using a Gd(III)-doped PA molecules and MRI imaging every five days. These scaffolds displayed characteristic degradation rates *in vivo* matching the time course of tissue regeneration ($t_{1/2}$ = ~14 days).

We transplanted FACS-isolated Î±7-integrin$^+$ CD34$^+$ muscle stem cells from *GFP/Luciferase* transgenic mice mixed with a 1wt%/vol aPA solution and then immediately injected into hindlimb muscles of pre-irradiated immunodeficient *NOD/scid* mice. When transplanted within aPA scaffolds, we observed profound improvements in the frequency of MuSC engraftment, as measured by bioluminescence imaging longitudinally up to one-month post-transplant, compared to MuSCs in control solutions. We also observed enhanced donor cell-derived myofiber contributions, as assayed by anti-GFP immunohistochemistry, a standard measure of muscle repair.

This work establishes a new stem cell encapsulation technology to aid muscle cell therapies. Importantly, this scaffold system self-assembles without the need for *in vivo* polymerization or cross-linking and has emergent nanofiber alignment to template muscle cell maturation. Simple rearrangement of PA amino acid sequence controls scaffold stiffness and can be varied to optimize muscle stem cell function *in vitro* and *in vivo*.

Blood Outgrowth Endothelial Cells (BOEC) Developed in Xeno-Free Conditions for Ischemic Tissue Regeneration

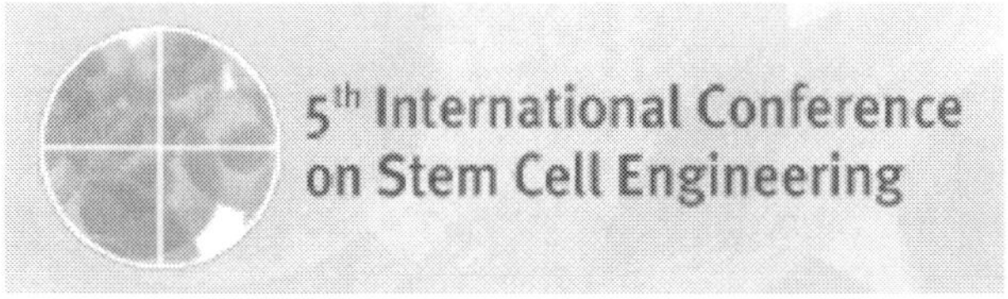

- Conference: International Conference on Stem Cell Engineering
- Year: 2016
- Proceeding: International Conference on Stem Cell Engineering 2016
- Group: Poster Submissions
- Session:

  Poster session

- Time:

  Tuesday, October 25, 2016 - 5:30pm-7:30pm

Authors:
Gratch, Y.

Ischemic tissues that develop as a result of various types of injury, often lack oxygen and nutrients required for both structural and functional regeneration. Inducing the emergence of an adequate vasculature (revascularization) can improve blood flow; allowing essential nutrients to re-perfuse the injured tissue. Transplantation of endothelial cells, either alone or with other supporting cell types, are a promising approach for augmenting recovery.

Blood outgrowth endothelial cells (BOEC) are a potentially autologous, highly proliferative, and readily available endothelial cell source. BOEC have shown similar potency in degree of vascularization when compared to the clinically-restricted human umbilical vein endothelial cells (HUVEC) both in vitro and in vivo in the presence of adult adipose-derived mesenchymal stromal cells (adMSC). Previously, we described a young source of immunoprivileged pericyte-like MSC, first trimester human umbilical cord perivascular cells (FTM HUCPVCs), that supports angiogenesis both in vitro and in vivo. We have developed xeno-free culture conditions using cGMP-compliant human platelet lysate (HPL) that preserves the phenotype (CD31+, CD144+, CD105+, CD45-)

and tube formation potential of BOEC in vitro. Our co-culture experiments have shown that, in vitro, BOEC and FTM HUCPVCs can interact on basal membrane extract (Matrigel^{TM}) to form tubes and endothelial cell sheets.

Our results suggest that clinically compatible angiogenic cell sources BOEC and FTM HUCPVC can be expanded in cGMP-compliant culture conditions for regenerative therapy after ischemic injury.

*** Formatting Issue - Best Available Paper/Graphic ***

Clinical-Grade Stem-Cell Based Model Systems for Mitochondrial Diseases

5th International Conference on Stem Cell Engineering

- Conference: International Conference on Stem Cell Engineering
- Year: 2016
- Proceeding: International Conference on Stem Cell Engineering 2016
- Group: Poster Submissions
- Session:

  Poster session

- Time:

  Tuesday, October 25, 2016 - 5:30pm-7:30pm

Authors:
Rao, R.
Grace, H.
Galdun, P. III
Serrano, M.
Buck, G.
Lesnefsky, E.
West, F.
Iyer, S.

Mutations in the mitochondrial DNA (mtDNA) lead to many human mitochondrial disorders with impaired bioenergetics. The clinical severity of the disease may also depend on the load of mutant mtDNA and bioenergetic dysfunction. The development of FDA â□□approved therapies for primary mitochondrial disorders has been hindered by genetic and clinical heterogeneity, limited knowledge of natural history of disease progression, and lack of clear clinical endpoints. Two recent studies have reported the derivation of human induced pluripotent stem cells (hiPSCs) using viral methods from two human diseased fibroblast cell lines carrying different mtDNA mutations. These studies demonstrated that reprogramming led to the generation of isogenic hiPSC clones; each containing different mutant mtDNA levels despite identical nuclear genomes. An important finding was that cell fate decisions in the hiPSCs were directly linked to the extent of mutant mtDNA burden. To further advance the search for improved cell models and therapies, we have used a highly efficient mRNA nuclear reprogramming technology, considered as the fastest and safest approach for generating integration-free, virus-free, clinical grade human induced pluripotent stem cells (hiPSCs). We are focusing our efforts on Leighâ s Syndrome (LS), as there is no current cure for LS nor an adequate model for understanding the rapid fatality associated with the disease.Â Our results indicate that (a) successful reprogramming of a LS fibroblast (carrying 8993T>G) was achieved (b) LS-hiPSC exhibited multiple transcription factors and cell surface markers representative of a pluripotent stem cell; (c) LS-hiPSC exhibited ability to differentiate into all three embryonic germ layers (d) LS-hiPSCs and differentiated derivatives continue to exhibit the 8993T>G mutation. Our results demonstrate the generation and characterization of a stable clinical-grade LS-hiPSC cell based model that now provide us the opportunity to investigate the effects of pathogenic mtDNA burdens on altered bioenergetics; and to monitor and quantitate mtDNA dynamics during differentiation into specialized cell types.

Combined Sry-Related HMG Box-4 (SOX4) and SOX11 Signaling in Nephron Progenitor Cells Is Required for Normal Renal Development In Vivo

5th International Conference on Stem Cell Engineering

- Conference: International Conference on Stem Cell Engineering
- Year: 2016
- Proceeding: International Conference on Stem Cell Engineering 2016
- Group: Poster Submissions
- Session:

  Poster session

- Time:

  Tuesday, October 25, 2016 - 5:30pm-7:30pm

Authors:
Hartwig, S.

Title: Combined Sry-Related HMG Box-4 (SOX4) and SOX11 signaling is required for normal renal development *in vivo*

Sushant Gavhale, Michel Arsenault, Blanca Esparza-Nino, Ashley Patriquen, Glenda M Wright, Sunny Hartwig.

Congenital anomalies of the kidney and urinary tract (CAKUT) are the most common cause of childhood end-stage renal disease. Members of the *Sry*-Related HMG Box-C (SOX-C) subfamily of nuclear transcription factors, including Sox4 and the closely-related Sox11, play overlapping and essential roles in controlling progenitor stem cell fate during organogenesis *in vivo.* We have previously demonstrated that both *Sox4* and *Sox11* are expressed in multiple progenitor cell populations during murine kidney development. Genetic ablation of *Sox4* in nephron progenitor cells ($Sox4^{nephron-}$ mice) results in CAKUT and end stage renal failure by 5 months *in vivo,* demonstrating an essential and non-overlapping role for *Sox4* in kidney development *in vivo.* Here we present our preliminary findings following conditional ablation of both *Sox4* and *Sox11* in the nephron progenitor lineage (*Sox4/11*$^{nephron-}$). In contrast to *Sox4*$^{nephron-}$ mice, *Sox4/11*$^{nephron-}$ animals die neonatally by postnatal day 1 (P1); lethality is associated with renal insufficiency (n=30) of variable presentation including bilateral renal agenesis (n=4), ulilateral renal agenesis (n=2), severe hypoglomerularopathy (n=7), and cystic renal dysplasia(n=4). Characterization of adult compound *Sox4/Sox11* heterozygotes is currently underway, along with quantitative assessment of the embryonic phenotypes of *Sox4/Sox11*$^{nephron-}$ kidneys, including quantitative assessment of renal morphogenesis, nephron progenitor proliferation, apoptosis and differentiation, nephrogenesis and marker analysis. Collectively, our work will define a novel *Sox4/11* signaling pathway that controls normal kidney development *in vivo*.

Acknowledgment: Funding from CIHR, Kidney Research Scientist Core Education and National Training (KRESCENT) Program and the Kidney Foundation of Canada (Sunny Hartwigs grant) are gratefully acknowledged.

*** Formatting Issue - Best Available Paper/Graphic ***

Controlling Chemical and Mechanical Cues to Augment Endothelial Fate from Human Induced Pluripotent Stem Cells

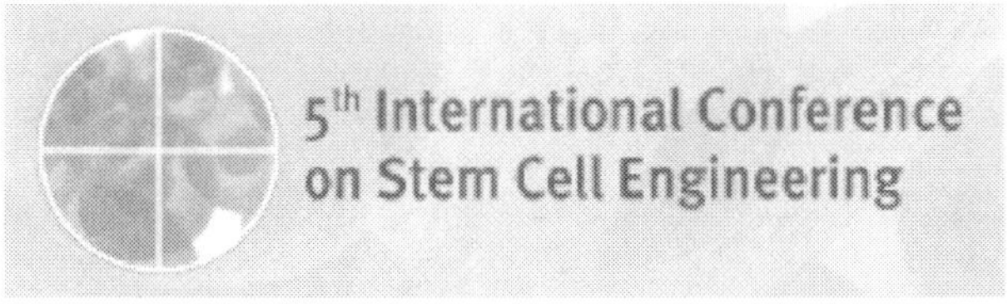

- Conference: International Conference on Stem Cell Engineering
- Year: 2016
- Proceeding: International Conference on Stem Cell Engineering 2016
- Group: Poster Submissions
- Session:

  Poster session

- Time:

  Tuesday, October 25, 2016 - 5:30pm-7:30pm

Authors:
Smith, Q., Johns Hopkins University
Chan, X. Y., Johns Hopkins University
Carmo, A., Tecnico Lisboa
Saunders, M.
Trempel, M., Johns Hopkins University

The efficient differentiation of functional endothelial cells (ECs) from human induced pluripotent stem cells (hiPSCs) is pivotal for their therapeutic potential. Aims to increase differentiation efficiency relies on the presentation of physiologically relevant cues including controlling the mechanical micro-environment. While several studies have demonstrated the effects of substrate stiffness on the differentiation efficiency of mesenchymal stem cells towards adipo/osteogeneic linages, less is known on the effects of substrate rigidity of hiPSCs differentiating towards vascular lineages.

Here we investigated whether mechanical dosing of hiPSCs during early stages of mesoderm induction biases endothelial differentiation.Â We identified the gene expression of the connective tissue growth factor (CTGF) and localization of the Yes-associated protein (YAP) when hiPSCs are cultured on polydimethylsiloxane (PDMS) (Youngâ□□s Modulus E ~ 3kPa or 1.7 MPa) or polystyrene substrates (E ~ 3 GPa). We found that mechanically dosing hiPSCs on 1.7 MPa substrates in mesoderm inducing media, increased Brachyury (T) expression when compared to culture on 3 kPa PDMS or tissue culture plates. In addition, when these mechanically dosed cells were continually differentiated towards a vascular phenotype, we found that cells primed on 1.7 MPa consistently lead to a robust endothelial phenotype. We conclude that differentiating hiPSCs are mechanosensitive through YAP and CTGF activation, leading to enhanced T expression during mesoderm induction, and robust endothelial differentiation.

Development and Functional Applications of Human iPSC-Derived Spinal Motor Neurons

5th International Conference on Stem Cell Engineering

- Conference: International Conference on Stem Cell Engineering
- Year: 2016
- Proceeding: International Conference on Stem Cell Engineering 2016
- Group: Poster Submissions
- Session:

  Poster session

- Time:

  Tuesday, October 25, 2016 - 5:30pm-7:30pm

Authors:
McMahon, C., Cellular Dynamics International
Wang, W. B., Cellular Dynamics International
Liu, J.
Chavez, C.
Meline, B.
Jones, E.
Kim, K. H.
Essex, A.
Martin, E.

The aim of this study was to produce spinal motor neurons from human iPSCs with sufficient purity for use in a multitude of downstream assays. In particular we wanted to produce motor neurons that could be cultured in defined conditions over long periods of time, without being hampered by outgrowth from proliferative cell types. Using an optimized differentiation protocol that improves upon published methods, we were able to produce motor neurons from iPSCs at greater than 60% purity as measured by Isl 1/2 and Tuj1 positive staining. These cells can be stored frozen, thawed, and cultured in media without glia for extended periods. We collected ICC, qPCR and MEA data to characterize the motor neuron cells and used iPSC lines from multiple donors to demonstrate a robust protocol that produces motor neurons independent of donor iPSC line. These data show the characteristics and utilization of motor neurons produced from iPSC.

Effect of Mesenchymal STEM CELLS and Their Secretome on Renal Progenitors CELLS in MODEL LPS Nephrotoxicity.

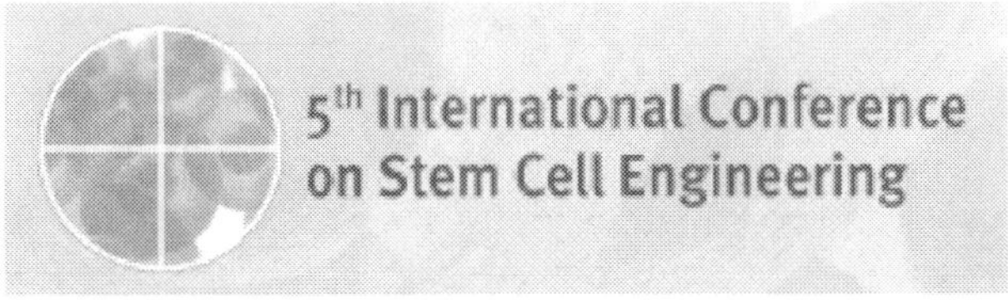

- Conference: International Conference on Stem Cell Engineering
- Year: 2016
- Proceeding: International Conference on Stem Cell Engineering 2016
- Group: Poster Submissions
- Session:

  Poster session

- Time:

  Tuesday, October 25, 2016 - 5:30pm-7:30pm

Authors:
Reis, L. A.
Keppeke, G. D.
Borges, F. T.
Simoes, M. D. J.
Schor, N.
We investigated the effects of mesenchymal stem cells (MSCs), your conditioned medium (CM) or exosomes (EXOs) in a LPS-induced nephrotoxicity and the role of renal stem progenitor cells (RPCs) in this model. Rats received i.v.: LPS (10 mg/Kg B.W.) or PBS (CTL) with MSCs (1x106), CM (500 Âµl) or its EXOs (100 Âµg/ml) from MSCs incubated or not for 12 hours with cytochalasin B (CB; 1 ÂµM) or actinomycin D (AD; 2.6 ÂµM), given in 1 or 3 administrations, sacrificed after 72 hours. Blood, urine samples were collected for creatinine (sCr), urea (sU) and FENa. Kidneys were analyzed for HE, KI67, caspase 3. EXOs as CD63and TEM. Y chromosome, IL6, TNF-Î±, INF-Î³ and IL10 were evaluated and RPCs for Wnt1, BrDU, PAX2, CD24. It was observed increases in sCr, sU, FENa, caspase 3 marking, proinflammatory cytokines and reduction of KI67 with lesions in proximal tubules by LPS. These parameters were ameliorated with MSCs, CM or EXOs. LPS and LPS+EXOs increased BrDU, Wnt1, PAX2, CD24 and CD63

expressions indicating positive mark of RPCs. CB and AD inhibited the protective effect of EXOs. The effect of 3 administrations of MSCs or CM or EXOs decreased mortality in LPS. Therefore, results support that the MSCs and its CM and EXOs protected from LPS-nephrotoxicity. It is reasonable to suggest that the mediation by EXOs at least in part, by expression RPCs in this cascade of events and finally, those EXOs alone could be employed in order to ameliorate LPS acute kidney injury (AKI).

Funds from: FAPESP, CNPq, CAPES and FOR

First Trimester Cord Perivascular Cells (FTM HUCPVCs) Home to Injured Cardiomyocytes and Promote MMP-Mediated Extracellular Remodeling Post Injury in Vitro and In Vivo Post Myocardial Infarct

5th International Conference on Stem Cell Engineering

- Conference: International Conference on Stem Cell Engineering
- Year: 2016
- Proceeding: International Conference on Stem Cell Engineering 2016
- Group: Poster Submissions
- Session:

  Poster session

- Time:

  Tuesday, October 25, 2016 - 5:30pm-7:30pm

Authors:
Iqbal, F., Create Fertility Inc.

Background: Following myocardial infarction, the reparative phase involves extracellular matrix (ECM) remodeling to replace necrotic myocytes with a collagen scar. An ideal therapeutic cell type should home to injury sites and effectively remodel ECM at injury sites. We aimed to investigate the homing properties of first trimester umbilical cord perivascular cells (FTM HUCPVCs) and MMP-based ECM remodelling using in vitro assays and in vivo, using a rat model of MI.

Methods: To assess homing properties of FTM HUCPVCs in vitro, primary rat cardiomyocytes were subjected to hypoxia (pO2=1%) and glucose deprivation in culture for 24h followed by labeled-FTM HUCPVCs seeded on MatrigelTM-coated trans-membrane inserts. Migration of FTM HUCPVCs was detected by florescent well scans. Neutralizing antibodies against SDF1, TNFÎ±, MMP2 and MMP15 were used to inhibit MatrigelTM processing and chemotaxis of MSCs. In vivo model of myocardial infarction

was established by left ventricular coronary artery ligation on 8 week old immunocompromised (Foxrnu) rats. 1 week post-MI, FTM-HUCPVCs, term-HUCPVCs or human BMSCs (n=6 per group) were injected into the injured myocardium (3X106 cells each). 2 weeks after cell implantation, animals were sacrificed. In situ gelatinase activity was assessed by fluorescent gelatinase substrate (DQ gelatin) on fixed cardiac tissue sections and the signal was quantified. 9 independent fields in 6 tissue sections per heart were analyzed. Statistical analysis was performed with ANOVA.
Results: MMP2 an MMP15 neutralizing antibodies significantly inhibited invasion of FTM-HUCPVCs towards injured cardiomyocytes in vitro ($p<0.05$). Serum-free culture conditions altered the trans-membrane migration of FTM-HUCPVCs in response to injured cardiomyocytes or soluble SDF1. Anti-TNFÎ± antibody significantly inhibited trans-membrane migration of FTM-HUCPVCs expanded in either serum-containing ($p<0.05$) or serum-free ($p<0.01$) conditions. Purified TNF-Î± and SDF1 attracted FTM-HUCPVCs through basal membrane-like structures, however the effect was not significant. FTM-HUCPVC-treated scar tissue displayed significantly increased gelatinase signal when compared to term-HUCPVCs and bone marrow MSCs.
Conclusion: ECM processing and homing in response to injured cardiomyocytes appeared to be mediated by MMPs as well as cytokines, including SDF1 and TNFÎ±, and altered by xeno-free conditions. These findings suggest that FTM-HUCPVCs could be an excellent candidate for post-MI cell therapy, and that the extent of MMP activity can be controlled by culture conditions. Our in vivo data also indicates that FTM-HUCPVCs have superior ECM processing in the scar tissue of infarcted hearts compared to older sources of MSCs.

*** Formatting Issue - Best Available Paper/Graphic ***

First Trimester Human Umbilical Cord Perivascular Cells Differentiated into Immunoprivileged, Contracting Cardiomyocyte-like Cells in Vitro

5th International Conference on Stem Cell Engineering

- Conference: International Conference on Stem Cell Engineering
- Year: 2016
- Proceeding: International Conference on Stem Cell Engineering 2016
- Group: Poster Submissions
- Session:

  Poster session

- Time:

  Tuesday, October 25, 2016 - 5:30pm-7:30pm

Authors:
Szaraz, P., University of Toronto

Myocardial infarction (MI) is the leading cause of mortality and morbidity worldwide. The extensive loss of cardiomyocutes without replacing them with healthy cells leads to congestive heart disease (CAD). Mesenchymal stromal cells are extensively studied for both allogeneic and autologous tissue regeneration applications due to their availability and beneficial immunomodulatory profile. While MSC based cell therapy is a widely investigated treatment option to substitute highly invasive surgical interventions, their limited differentiation potential restricts applications aiming to replace host cardiomyocytes. The ideal MSC type should have the ability to provide micro- or macro-tissue significant to achieve cardiac regeneration after MI.

We aimed to investigate the cardiomyogenic differentiation potential of first trimester human umbilical cord perivascular cells (FTM HUCPVCs), a novel, young source of immunoprivileged mesenchymal stromal cells. Our aggregation-based differentiation method achieved significantly increased expression of cardiomyocyte markers (cTnT, MYH6, SIRPA, and CX43) in both FTM and term HUCPVCs compared to bone marrow

MSCs, while HUCPVCs immunogenicity remained significantly lower as indicated by HLA-A and HLA-G expression levels and significantly lower susceptibility towards T cell mediated cytotoxicity.

FTM HUCPVCs showed increased aggregate formation potential and generated spontaneously contracting cells within 1 week of co-culture with rat cardiac feeder layer. This makes them the first MSC type that displays such ability. Our results indicate that young FTM HUCPVCs have superior cardiomyogenic potential combined with immunopriviledged properties when compared to MSCs of older tissue sources. We postulate that in vitro pre-differentiation of MSCs from young sources â□□ such as FTM HUCPVCs - could be a potential strategy to increase their efficiency in vivo and provide with better treatment strategies for MI.

Geometric Cues in Bioengineered Scaffold Enhance Podocyte Differentiation in Vitro

5th International Conference on Stem Cell Engineering

- Conference: International Conference on Stem Cell Engineering
- Year: 2016
- Proceeding: International Conference on Stem Cell Engineering 2016
- Group: Poster Submissions
- Session:

  Poster session

- Time:

  Tuesday, October 25, 2016 - 5:30pm-7:30pm

Authors:
Korolj, A., University of Toronto
Radisic, M., University of Toronto
Zhang, B., University of Toronto
Laschinger, C., University of Toronto
James, C. H.
Hu, E.
Willette, R. N.

Title: Geometric cues in bioengineered scaffold enhance podocyte differentiation *in vitro*

Session: Differentiation/morphogenesis

Words: 484

This work reports on a novel, biomimetic, microfabricated platform that has been engineered to induce functional differentiation of podocytes, a cell type found in the kidney nephron.

The kidney nephron is the biological unit in which waste is filtered from the blood into the urine. In the nephrons glomerulus, podocytes wrap around a dense cluster of capillaries and along with the glomerular basement membrane and fenestrated endothelium, forms the functional filtration unit. Podocytes have an arborized morphology with interdigitated processes that wrap around the capillaries and connect to other podocytes. It has been demonstrated that the slit diaphragm on these processes, marked by the gene nephrin, is the barrier structure for filtration in the glomerulus[1], and a hallmark of differentiation.

The *in vitro* study of these cells is driven by a need for mechanistic understanding of podocyte barrier function, which is an essential common pathway in proteinuric diseases of the kidney. Many kidney diseases are associated with dedifferentiation/dysfunctions of podocytes, where they lose the specialized features required for their function. Thus, the study of podocytes *in vitro* is focused on understanding the cells biology and the hallmarks of differentiation, which has been done by exploring the chemical environment, co-cultures and perfusion conditions. These studies are currently conducted on plated cells in a regular two-dimensional (2D) culture, where it is difficult to reach appreciable levels of nephrin expression. A cell culture system that incorporates a 3D physical environment has not yet been investigated.

In the glomerulus *in vivo*, the podocyte experiences out-of-plane curvature. We hypothesized that a biomimetic topography, that mimics the curving glomerular capillary structures, would provide a physical stimulus that promotes podocyte differentiation *in vitro*, leading to more appreciable levels of nephrin expression.

We engineered a culture platform that resembles the rounded and densely arranged capillary structures of the glomerulus by using spherical glass beads in a modified microfabrication technique. 100 m diameter glass beads were cured onto a silicon wafer to create a mold for a platform that has out-of-plane curvature as its topography. This topography presents cells with curvature akin to curved glomerular capillaries.

Podocytes were cultured on the topographic platform and compared to a flat platform. After 9 days of differentiation, cell samples were either fixed for imaging, or RNA was collected for reverse transcription and subsequent qPCR to quantify gene expression. In all experiments, the topographic platform consistently yielded cells with more arborized morphologies and greater upregulation of nephrin, than in the flat platform. While the trend was present in experiments with non-supplemented culture media, it was more pronounced where cells were cultured with a hormone and vitamin supplement.

This work presents a new parameter, that of 3D physical environment, to consider and incorporate into the culture of podocytes for improved differentiation. It shows the effectiveness of biomimicry toward the engineering of higher-fidelity cells, and contributes to the development of physiologically relevant biological models for the study of pathology.

High Content Drug Testing Platform Using Matured Human Engineered Cardiac Tissue

5th International Conference on Stem Cell Engineering

- Conference: International Conference on Stem Cell Engineering
- Year: 2016
- Proceeding: International Conference on Stem Cell Engineering 2016
- Group: Poster Submissions
- Session:

  Poster session

- Time:

  Tuesday, October 25, 2016 - 5:30pm-7:30pm

Authors:
Zhao, Y., University of Toronto
Radisic, M., University of Toronto
Feric, N., University of Toronto
Zhang, B., University of Toronto
Aschar-Sobbi, R.
Ronaldson, K., Columbia university
Conant, G., University of Toronto
Panhke, A., University of Toronto
Vunjak-Novakovic, G., Columbia University
Backx, P.

The current drug development process is both very slow (15 year average) and costly ($1.5B/drug average). Despite this hefty investment, drugs are routinely withdrawn from the market due to serious toxicities and adverse cardiovascular effects, indicating a failure of the current drug screening methods to fully detect and characterize drug effectiveness and toxicity on human cardiomyocytes (CMs). Safety screen assessments performed on cell cultures and animal models do have predictive value but the results obtained may not correlate with clinical risk. Specifically, this entire framework predicts only 75-90% of the effects of a drug in a human patient with the result that many promising drugs are abandoned early in development and some cardiotoxic drugs go to market. A potential explanation is that much of the safety and efficacy testing has relied upon animal studies and little information has been gathered from human preparations. The applicability of animal studies is limited by the fundamental cardiovascular differences amid species. Similarly, cardiac ion channel *in vitro* screens frequently involve non-cardiac cells (Chinese hamster ovary, CHO or human embryonic kidney, HEK) that have been engineered to overexpress an individual target ion channel, most commonly the human ether-a-go-go (hERG) channel (I_{Kr} current). However, interference with other currents or currents in combination could also contribute to cardiac complications necessitating an assessment protocol that evaluates the effect of a drug on the collective behaviour of ion channels in a cardiac tissue.

To address these limitations we have developed the Biowire II platform, a human cardiac tissue array using CMs derived from human induced pluripotent stem cells (hiPSCs) that: 1) captures the physiological hallmarks of the adult human myocardium, 2) is situated in an inert plastic platform compatible with current drug screening equipment and practices in the pharmaceutical industry and 3) enables on-line non-destructive readouts of contractile force and Ca^{2+} transients, as well as the collective behaviour of ion channels.

In the Biowire II platform, immature human CMs and support cells in a hydrogel matrix were seeded into inert polystyrene microwells fitted with a pair of flexible polymer wires for tissue alignment. Upon compaction, the human cardiac tissues were electrically field stimulated over the course of weeks and were demonstrated to have highly organized myofibril ultrastructure, improved electrical properties (ET, MCR), a positive force-frequency relationship (1-4Hz), post-rest potentiation, a recorded conduction velocity of 35.7 cm/s and an action potential profile with human adult-like characteristics (I_{to1} notch, plateau, a rapid upstroke velocity and resting membrane potential of -70mV). The Biowire II platform was also demonstrated to be amenable to various cell sources (CDI iCell CMs and CMs^2; human embryonic stem cell derived Hes2 and Hes3 CMs; hiPSC BJ1D and C2A CMs), and the engineered human cardiac tissues obtained using the Biowire II platform were both physically and functionally reproducible. Using non-destructive force readouts, the chronotropic and inotropic effects of isoproterenol, nifidipine and thapsigargin were captured. Whole cell patch clamp recordings from the intact Biowire II tissues demonstrated the effects of dofetilide, flecainide and 4-Aminopyridine on action potential. Taken together, the inert plastic Biowire II platform can generate high fidelity human engineered cardiac tissues that possess structural and functional maturation hallmarks of human adult myocardium and can be used as an *in vitro* drug testing platform to provide non-destructive contractile and calcium handling readouts as well as to interrogate the effect of drugs on the collective ion channel behaviour of human cardiac tissue.

High-Throughput Analysis of Kinase Inhibitor Drugs on Cardiac Function Using Engineered Heart Tissue Constructs

5th International Conference on Stem Cell Engineering

- Conference: International Conference on Stem Cell Engineering
- Year: 2016
- Proceeding: International Conference on Stem Cell Engineering 2016
- Group: Poster Submissions
- Session:

  Poster session

- Time:

  Tuesday, October 25, 2016 - 5:30pm-7:30pm

Authors:
Conant, G., University of Toronto
Ahadian, S., University of Toronto
Zhao, Y., University of Toronto
Radisic, M., University of Toronto

A serious concern in health care today is the unintentional long-term effects of pharmaceuticals on the heart, which are not detected during clinical trials. Tyrosine kinase inhibitors (TKIs) are commonly used as highly specific cancer targeting agents for their ability to prevent tyrosine kinase molecules from activating signaling pathways that regulate cell growth, differentiation, metabolism, migration, and programmed cell death. Due to the highly metabolic nature of cardiac tissue, cardiomyocytes are sensitive to perturbations in mitochondrial function. Clinically used TKIs have been found to have a spectrum of toxicities on the heart that were not detected during clinical trials. High-throughput preclinical screening of TKIs could offer much information as to their cardiovascular effects and suggest ideal candidates for further investigation without putting patients at risk. We hypothesized that a high-throughput, high-fidelity screening platform for evaluating the physiological effects of TKIs on human myocardium could be established using high-content screening assays, artificial neural network (ANN) modelling, and engineered heart tissue constructs termed Biowires.

High-content screening assays from Molecular Devices were used sequentially to assess cell viability and basic cardiac function of 80 TKIs on a human cardiomyocyte monolayer at 3 different concentrations. Experimental results were then analyzed using ANN modelling. A network was designed that could predict the compound with the least detrimental effects on cardiac function. This compound was then examined in the Biowire II platform, a micro-scale cardiac tissue array mimicking physiological tissue that consist of human cardiomyocytes seeded onto polymer wires and electrically stimulated in order to promote tissue maturation. Fluorescent image microscopy was used to examine tissue contraction and calcium transient profiles.

Initial monolayer testing using human induced pluripotent stem cell-derived cardiomyocytes (iPS-CM) and 2 high-content screening assays yielded four parameters for each inhibitor at each concentration: cell viability, live cell number, calcium peak magnitude, and contraction frequency. ANN modelling was used to analyze the experimental results efficiently and in an unbiased manner, to reduce the number of TKIs tested on the Biowire platform. We chose ANN modelling because of its predictive capabilities and its ability to identify complex non-linear relationships between molecular function and cell response. The effect of the least detrimental TKI was then examined on matured cardiac tissue to observe changes in force of contraction and calcium transients after exposure, as well as immunohistochemical effects. In the future, these inhibitors of interest could be studied to understand their cellular mode of action.

In conclusion, this system linked high throughput experimental screening with an empirical modelling tool to observe and measure the physiological effects of TKIs on cardiac tissue. We believe this system can be applied to screening of other compounds on cardiac tissue and should be used to minimize the cost of experimentation and maximize the impact of pre-clinical research.

Hypoxia-Inducible Hydrogels Facilitate Study of Cluster-Based Neovascularization in Vitro

5th International Conference on Stem Cell Engineering

- Conference: International Conference on Stem Cell Engineering
- Year: 2016
- Proceeding: International Conference on Stem Cell Engineering 2016
- Group: Poster Submissions
- Session:

  Poster session

- Time:

  Tuesday, October 25, 2016 - 5:30pm-7:30pm

Authors:
Blatchley, M., Johns Hopkins University
Wang, S., Johns Hopkins University
Hall, F. III, Mississippi State University
Hypoxia-inducible hydrogels facilitate study of cluster-based neovascularization *in vitro*

Michael R. Blatchley1,2, Songnan Wang2, Franklyn Hall3, Sharon Gerecht2,4

1Department of Biomedical Engineering, Johns Hopkins University School of Medicine, Baltimore, MD, USA;

2Department of Chemical and Biomolecular Engineering and Institute for NanoBioTechnology , Johns Hopkins University Whiting School of Engineering, Baltimore, MD, USA, 3Department of Chemical Engineering, Mississippi State University, Starkville, MS, USA, 4Department of Materials Science and Engineering, Johns Hopkins University Whiting School of Engineering, Baltimore, MD, USA

The mechanisms governing postnatal vasculogenesis have begun to be uncovered in the last several decades. These mechanisms have informed the design of numerous therapeutics, seeking to regenerate or control the formation of new vasculature following loss-of-function diseases, such as peripheral and coronary artery diseases, as well as provide an understanding of tumor angiogenesis. The processes which govern vascular regeneration in such clinical pathologies involve complex spatiotemporal interplay between physical and biochemical cues. The plenary work of several groups has described one mechanism by which blood vessels may form *de novo*. In this body of work, new blood vessels are formed through intracellular changes in a population wide phenomenon, that subsequently leads to sprouting and branching through coalescence of single cells to form multicellular, luminal vessels. In addition to these works, observational reports have indicated the presence of a secondary, potentially parallel mechanism. In particu lar, endothelial progenitor cell (EPC) cluster formation has been reported in ischemic and hypoxic tissues. This mechanism commences with the formation of multicellular clusters that then sprout to form luminal structures, which can anastomose with existing vasculature. However, an understanding of the regulatory mechanism governing these processes is limited, due to a shortage in model systems that recapitulate the pathological microenvironmental cues.

The advent of biomaterials technologies has enabled precise control over *in vitro* microenvironments. Such biomaterials have facilitated studies which have uncovered a variety of biomechanical and biochemical cues that influence tissue morphogenesis. Specifically, we have developed hydrogel materials with precise control over two potent regulators of vascular regeneration: oxygen tension and matrix viscoelasticity. Our previous works established hydrogels with O2 gradients ranging from <1% to 21%. However, cellular responses to hypoxia vary widely based on specific and precise changes in O2 tension. Here, by generating layered hydrogels that discretize O2 gradients, we have studied cellular behavior and examined the effects of varying O2 tensions in 3D. Additionally, we can predictably tune matrix mechanics within each layer by the addition or omission of a secondary crosslinker, microbial transglutaminase (mTG).

Our initial observations indicate that hypoxic cluster-based vasculogenesis comprises a two-step mechanism: (1) cluster formation, stabilization and expansion, and (2) cluster sprouting and branching. Interestingly, we have consistently observed that low O2 tension promotes clustering of EPCs. We have also shown that this process is mediated by matrix metalloproteinase (MMP) production, which leads to matrix degradation that enables cell clustering. Upon concurrent encapsulation of EPCs with DQ-gelatin, which fluoresces in the presence of MMPs, we have confirmed the role of MMP in cluster formation. Additionally, we have analyzed junctional and cytoskeletal protein expression including intercellular adhesion molecule 1 (ICAM-1), integrin Î²2, vascular endothelial cadherin (VE-cad), and phalloidin, to delineate how cell-cell interactions mediate cluster formation, maintenance, and expansion. To determine the important factors regulating the second step of the proposed process, we have specifically and dynamically controlled matrix viscoelasticity. Dynamic regulation of matrix mechanics, specifically increases in matrix stiffness and crosslinking density by addition of mTG, increased sprouting from EPC clusters by coaxing cell-ECM interactions.

Intracellular Oxidation in Murine Stem Cells Due to O2 and Glucose Culture Conditions

5th International Conference on Stem Cell Engineering

- Conference: International Conference on Stem Cell Engineering
- Year: 2016
- Proceeding: International Conference on Stem Cell Engineering 2016
- Group: Poster Submissions
- Session:

  Poster session

- Time:

  Tuesday, October 25, 2016 - 5:30pm-7:30pm

Authors:
Seals, S.
Pan, C., Georgia Institute of Technology & Emory University
Aigen, R.

Clinical realization of stem cell therapies requires more knowledge at the mechanistic level to control differentiation or maintain pluripotency to yield cellular products with low heterogeneity. Recent characterization of metabolism at different stages of pluripotency suggests that directing embryonic stem cells (ESCs) towards specific phenotypes can be augmented through the use of nutrient and oxygen culture conditions; however, it is unclear how these metabolic conditions synergize with growth factor supplementation and media inhibitors for ESC expansion or differentiation. In particular, a reactive oxygen species (ROS) byproduct of hypoxia and oxidative phosphorylation, H_2O_2, is a known modulator of MAPK and SMAD signal transduction cascades and thus may affect cell differentiation decisions. We hypothesize that hypoxia inducible factor (HIF) activated ROS allows modulation of signaling pathways, thereby allowing one to metabolically tune cellular phenotype using culture conditions in a reproducible manner. In order to test this hypothesis, we are quantifying the relative magnitude and variance of intracellular H_2O_2 levels in mESC and mEpiSC colonies under a range of oxygen and glucose conditions. The epiblast cell line is used as a comparison to hESCs because both use an Activin A/FGF-dependent mechanism for pluripotency maintenance, which is significantly different from the BMP/LIF-dependent mechanism in mESCs. A ratiometric sensor protein, HyPer, that provides quantitative intracellular measurements specific to H_2O_2, was stably transfected in the mESC-D3 cell line to provide unprecedented quantification of H_2O_2. The D3-HyPer cells were cultured in monolayer for three days under oxygen and glucose conditions relevant to diffusion limitations in vivo (20% and 5% O_2; 25 mM and 5.5 mM glucose-supplemented media). Colonies were imaged by live cell confocal microscopy and analyzed for ratiometric (oxidized/reduced) value and variance of cytosolic signal among cells. The mESC HyPer-D3 cell line showed statistically significant ($p < 0.0001$) increased ratios of oxidized/reduced HyPer for the low glucose conditions when compared to high glucose, regardless of the oxygen environment during culture. The ratio increased by 1.7x for low glucose/high oxygen when compared to high glucose/high oxygen and 1.5x for low glucose/low oxygen when compared to high glucose/low oxygen. Statistically significant differences ($p < 0.0001$) were observed when glucose supplementation was held constant and oxygen was the variable, but not to the same magnitude. This suggests that the highly glycolytic mESCs are more sensitive to glucose deprivation than hypoxia for eliciting cellular oxidation. Furthermore, despite reports of hypoxia-induced ROS in other cell types, no synergetic effects were noted in the low glucose/low oxygen condition. The mEpiSC HyPer-D3 cell line showed a statistically significant ($p < 0.0001$) increased ratiometric HyPer value for the low glucose/high oxygen condition when compared to all other conditions. Future studies will investigate how metabolic shifts between oxidative phosphorylation and glycolysis during pluripotent-to-epiblast and epiblast-to-germ layer transitions are reflected in the intracellular H_2O_2 levels and regulation pathways. Increased knowledge of how metabolism regulates signaling and pluripotent transcription factors may result in fundamental shifts in stem cell culturing protocols for expansion at the pluripotent state or directed towards lineage specification in a reproducible and homogeneous manner.

Intracellular Reactive Oxygen Species Associated with Embryonic Stem Cell Differentiation

5th International Conference on Stem Cell Engineering

- Conference: International Conference on Stem Cell Engineering
- Year: 2016
- Proceeding: International Conference on Stem Cell Engineering 2016
- Group: Poster Submissions
- Session:

  Poster session

- Time:

  Tuesday, October 25, 2016 - 5:30pm-7:30pm

Authors:
Pan, C., Georgia Institute of Technology & Emory University
Fan, Y., Georgia Institute of Technology

Reactive Oxygen Species (ROS) generated from mitochondrial respiration and other metabolic processes are important signaling molecules and have been linked to stem cell differentiation. Embryonic stem cells (ESCs) have the capacity to give rise to almost all cell lineages in the body, serving as a great system to study the mechanisms regulating cellular differentiation. ESCs maintain low ROS levels as compared with differentiated cells through glycolytic metabolism and enhanced antioxidant defense capability. However, how the redox status affects lineage specification during ESC differentiation remains elusive. Here, we have established a biosensor system to monitor the intracellular hydrogen peroxide (H_2O_2) level, a major form of ROS, in ESCs and cell aggregates. In this system, the cytoplasmic protein HyPer is stably expressed in ESCs and displays different spectral properties in presence of reducing and oxidizing agents. Therefore, ratiometric analysis of oxidized and reduced HyPer signals can be performed for individual cells in fluorescence microscopy images and the oxidized/reduced HyPer (Oxi/Red) ratio can serve as a sensitive indicator of the intracellular H_2O_2. We differentiated ESCs toward neural lineages via embryoid body formation and treatment with retinoic acid for 3 days followed by a neural differentiation regimen for another 5 days. Our results show that ~3-fold dynamic range in the Oxi/Red ratio for fully oxidized (1.31±0.47, hydrogen peroxide treated) vs. fully reduced HyPer signal (0.51±0.18, dithiothreitol treated, $p < 0.0001$). The baseline H_2O_2 level in ESCs, with an Oxi/Red ratio of 0.46(±0.14), is lower than reported in the literature for any other cell type, reflecting a highly reduced intracellular environment. In contrast, neural lineages derived from ESCs show stronger oxidized HyPer signal and weaker reduced signal, resulting in a statistically significant increase in the intracellular H_2O_2 level (Oxi/Red ratio = 0.79±0.17, $p < 0.0001$). The increase in H_2O_2 is consistent with known metabolic shifts to oxidative phosphorylation upon loss of pluripotency. ESCs also display a significant heterogeneity in intracellular H_2O_2, which may correlate with the differentiation potential of individual cells, as the oxidative response of the neural-derived cells demonstrated a smaller variance within the population. This unique imaging tool provides the opportunity to study spatiotemporal redox patterning within 3D cell aggregates and allows profiling the redox statuses in various cellular systems ranging from ESCs to differentiated cells. Future studies will provide significant insights into the lineage-specific redox regulation of stem cell differentiation.

*** Formatting Issue - Best Available Paper/Graphic ***

Investigating the Role of Hypoxia-Inducible Factor-1 Activation in the Vascularization of Modular Tissue Engineered Constructs

5th International Conference on Stem Cell Engineering

- Conference: International Conference on Stem Cell Engineering
- Year: 2016
- Proceeding: International Conference on Stem Cell Engineering 2016
- Group: Poster Submissions
- Session:

  Poster session

- Time:

  Tuesday, October 25, 2016 - 5:30pm-7:30pm

Authors:
Lam, G., University of Toronto
Sefton, M. V., University of Toronto

Vascularization is crucial to the fabrication of engineered tissues of therapeutically relevant size and cell loads. While prolonged hypoxia is detrimental to cell survival in these constructs, it is also a necessary driver of angiogenesis, exerting many of its effects via the hypoxia inducible factor (HIF)-1 pathway[1,2].

Modular tissue engineered constructs are formed by random assembly of micrometer-scale cylindrical collagen pieces (â□□modulesâ□□) embedded with adipose-derived mesenchymal stromal cells (adMSC) and enveloped with endothelial cells (EC); injected modules remodel *in vivo* to form a vascularized organoid[3,4]. In a recent study, varying cell densities, module diameter and implant volume affected graft-derived vascularization. Vessel densities were greatest ($p<0.01$) in implants with pronounced nuclear expression of HIF-1Î± at day 7, as with â□□largeâ□□ volume implants (0.10 mL, 1.5Ã□10^7 adMSC/mL, 3.9Ã□10^4 EC/mL) and â□□smallâ□□ volume implants with the highest cell density (0.01 mL, 1.5×10^8 adMSC/mL, 9.8×10^4 EC/mL)[5]. HIF-1 is hypothesized to govern: 1) the survival of graft-derived cells immediately following implantation, and 2) subsequent recruitment of angiogenic myeloid cells to the hypoxic site.

Here, the role of HIF-1 on module-induced vascularization was elucidated using pharmacological and genetic methods. SCID-bg mice were provided daily injections of 2 mg/kg digoxin or saline starting 2 days prior to implantation of modules (*in vitro* preconditioned in 100 nM digoxin or saline for 6 hours), and continuing on to 7 days post-implantation. Digoxin-treated implants showed reduced graft-derived and total vessel formation compared with saline-treated controls at day 7 (19±1.9 vs. 44±9.9 $UEA1^+$ graft-derived vessels/mm^2 and 19±3.0 vs. 46±10 $CD31^+$ total vessels/mm^2; $p<0.01$; n=5). Effects of HIF-1 inhibition correlated with reduced recruitment of endothelial progenitor cells ($CD45^-CD117^+$ $VEGFR2^+$) to module implants 3 days post-surgery (6.0Â±1.0 vs. 18Â±2.0 cells/mg implant; $p<0.01$; n=5).

In other experiments, HIF-1 was inhibited by lentiviral-mediated transduction of HUVEC with short hairpin RNA (shRNA) targeting HIF-1Î±. Upon implantation in modules, knockdown of HIF-1Î± in HUVEC reduced total vessel densities relative to targeting a scramble sequence at day 7 (13Â±4.0 vs. 26Â±4.0 $CD31^+$ total vessels/mm^2; $p<0.05$; n=5). However, no effect was seen at day 14, likely due to compensatory effects of HIF-2Î± upregulation at day 7 (135Â±8.0 vs. 54Â±13 HIF-2Î±$^+$ nuclei/mm^2; $p<0.01$; n=5).

Results from pharmacological and genetic methods of inhibition highlight HIF-1 as an early driver of module-induced vascularization, potentially exerting its effects by modulating recruitment of angiogenic EPC to the implant. Currently, we are examining ways of incorporating a HIF stabilizing small molecule, desferrioxamine, into modular tissue engineering to increase the rate of mature vessel formation. Knowledge gained from these studies may be harnessed to better apply modular tissue engineering to advance cell therapy.

References

1. Zhang, X. *et al.* (2011) Wound Repair Regen **18(2):** 193-201.
2. Semenza, G.L. (2014) Annu Rev Pathol **9:** 47-71.
3. McGuigan, A. and Sefton, M.V. (2006) PNAS **103(31)**: 11461-66.
4. Butler, M.J. and Sefton, M.V. (2012) Tissue Eng Part A **18(15- 16)**:1628-41.
5. Lam, G.C. and Sefton, M.V. (2015) Tissue Eng Part A **21(3-4)**: 803-16.

*** Formatting Issue - Best Available Paper/Graphic ***

Matrix Stiffness Induces Suppressor Cell Phenotype in CD14+ Bone Marrow mononuclear Cells

5th International Conference on Stem Cell Engineering

- Conference: International Conference on Stem Cell Engineering
- Year: 2016
- Proceeding: International Conference on Stem Cell Engineering 2016
- Group: Poster Submissions
- Session:

  Poster session

- Time:

  Tuesday, October 25, 2016 - 5:30pm-7:30pm

Authors:
Vining, K. H., Harvard University
Mooney, D. J., Harvard University

Myeloid-derived suppressor cells (MDSCs) suppress anti-tumor immune responses and contribute to tumor resistance to immunotherapy, but how their function is regulated by the tumor microenvironment is unclear. As tumors typically are stiffer than normal healthy tissue, we hypothesized that matrix mechanical properties play a significant role in the immune function of MDSCs. Tumor stiffness was modeled with a collagen-alginate interpenetrating network (IPN), and oscillatory shear rheology was performed to characterize the hydrogelsâ□□ resulting viscoelastic properties, i.e., storage and loss moduli and stress relaxation.Â We isolated primary human CD14+ mononuclear cells from bone marrow, whichÂ include a subset of myeloid progenitors that give rise to monocytes, macrophages, and dendritic cells in peripheral blood. The isolated cells were encapsulated in hydrogels with a range of composition and stiffness. qPCR analysis after three days of culture revealed that increased stiffness of the collagen-alginate IPN was associated with elevated expression of MDSC markers, including cyclooxygenase-2 (PTGS2), indoleamine 2,3-dioxygenase (IDO1), and IL6. Further, analysis of conditioned media revealed that stiffer hydrogels were associated with increased amounts of secreted prostaglandin E2 (PGE2), which is associated with tumorigenesis and immune-resistance in melanoma. Support for the physiologic relevance of these findings came from analysis of processed lysates from B16 melanoma tumors, which were found to contain detectable levels of prostaglandin E2 (PGE2). These findings indicate that CD14+ bone marrow mononuclear cells are responsive to extracellular matrix mechanics, and higher stiffness induces a suppressor cell phenotype. Thus, mechanical interactions between stroma and myeloid cells may have relevant implications in tumor resistance to cancer immunotherapies.

Myocardial Matrix Hydrogel Enhances Cardiac Spheroid Generation and Cardiogenic Commitment

5th International Conference on Stem Cell Engineering

- Conference: International Conference on Stem Cell Engineering
- Year: 2016
- Proceeding: International Conference on Stem Cell Engineering 2016
- Group: Poster Submissions
- Session:

  Poster session

- Time:

  Tuesday, October 25, 2016 - 5:30pm-7:30pm

Authors:
Gaetani, R., UCSD
Sluijter, J.
Christman, K., University of California, San Diego

Cardiac progenitor cells (CPCs), which have been examined for the treatment of myocardial infarction, have been shown to have a superior immunomodulatory and paracrine effect when in the form of spheroids compared to single cells. We previously showed that a myocardial matrix hydrogel enhanced cardiogenic properties of disassociated CPCs grown in 2D culture or when encapsulated in the hydrogel in 3D. The aim of this study was to evaluate the effects of the myocardial matrix when incorporated into cardiac spheroids. Sca-1+ CPCs were cultured in poly-D-lysine coated wells with or without the addition of myocardial matrix or collagen type I at 0.06 mg/ml. To form the myocardial matrix hydrogel, porcine left ventricular tissue was decellularized, digested in pepsin, and then diluted in media 1 day after cell seeding. Spheroid number and size were measured after 5 days in culture. A metabolic assay was performed at day 1, 3 and 5. Gene expression analysis of cardiac and vascular markers was evaluated after 4 days.

Myocardial matrix enhanced the generation and diameter of cardiac spheroids in culture compared to media only (113.2±25.9 and 60.0±14.8 μm respectively). Interestingly, no spheroids were observed when collagen was supplemented, but rather the cells clumped together forming a non-homogeneous large cell aggregate. An Alamar Blue assay showed a significant increase in CPC metabolic activity after 5 days in culture when cells were cultured in the presence of matrix. Similarly, gene expression analysis showed a significant increase of the cardiac transcription factor Nkx2.5 with trends towards an increase in GATA-4, Mef2c, and the sarcomeric markers Troponin I and cardiac actin. No differences in the expression of the vascular markers VEGF-R2 and CD31 were observed. Our study demonstrated that the incorporation of myocardial matrix into cardiac spheroids enhances their formation, size, metabolic activity, and cardiac commitment. Our approach represents a promising alternative and advantageous culture condition for cell-based cardiac regeneration or for tissue engineering self-assembly strategies.

Acknowledgments: This work was supported by the NHLBI (1R01HL113468).

Disclosures: KLC is co-founder, board member, consultant, and holds equity interest in Ventrix, Inc.

Optimized Media and Workflow for the Expansion of Human Pluripotent Stem Cells As Aggregates in Suspension Cultures

5th International Conference on Stem Cell Engineering

- Conference: International Conference on Stem Cell Engineering
- Year: 2016
- Proceeding: International Conference on Stem Cell Engineering 2016
- Group: Poster Submissions
- Session:

  Poster session

- Time:

  Tuesday, October 25, 2016 - 5:30pm-7:30pm

Authors:
Jervis, E., StemCell Technologies Inc
McLaughlin, A., STEMCELL Technologies Inc
Hukezalie, K., StemCell Technologies
Woodside, S., StemCell Technologies
Thomas, T. E., STEMCELL Technologies Inc
Eaves, A. C., British Columbia Cancer Agency
Louis, S. A., STEMCELL Technologies Inc

3D suspension culture enables the efficient and cost-effective scale-up of human pluripotent stem cell (hPSCs) manufacturing. However, the current practice of using media optimized for 2D adherent cultures directly in 3D suspension culture systems, can lead to low volumetric productivity and inefficient workflow. To overcome these limitations we developed mTeSR™3D, a defined medium based on mTeSR™1, and novel protocols for fed-batch culture of hPSC aggregates. Human embryonic stem cell (hESC) lines (H1 or H9) or human induced pluripotent stem cell (hiPSC) lines (WLS-1C or STiPS-M001) that were previously maintained in 2D mTeSR™1 culture were seeded into multiple suspension culture vessels (e. g. shaker bottles, spinner flasks and bioreactors) containing mTeSR™3D Seed Medium plus 10 Î¼M Y-27632 ROCK inhibitor. 3D cultures were maintained using either daily 50% mTeSR™1 medium exchanges (control) or using a fed-batch protocol whereby the culture medium was supplemented daily with mTeSR™3D Feed Medium. After 3 or 4 days in suspension culture, aggregates were harvested, dissociated into small clumps with Gentle Cell Dissociation Reagent (GCDR) or single cell suspensions enzymatically, and re-seeded in mTeSR™3D Seed Medium plus 10 Î¼M Y-27632. Passaging and feeding cycles were repeated for at least 5 passages. 3D cultures were assessed for growth, viability, hPSC marker expression, *in vitro* differentiation potential, and karyotype. In addition, media was analyzed for molar glucose to lactate yield to characterize metabolism. By day 4, aggregates cultured in mTeSR™3D typically grew to a mean diameter of 350 Î¼m, with a 5-fold increase in cell number. Image analysis was routinely performed to estimate aggregate size during growth. Using mTeSR™3D, up to 10^9 cells can be produced starting from a single 6-well plate within 2-3 weeks representing a greater than 500-fold expansion. hPSC cultures maintained in mTeSR™3D differentiated into all 3 germ layers with high efficiency. The average volumetric productivities were 0.7, 3.1 and 6.9 ($x10^5$) viable cells / mL in 2D, with daily 50% media exchange, and mTeSR™3D fed-batch protocols, respectively. Using the GCDR clump passaging protocol, mTeSR™3D cultured hPSCs retained normal karyotypes. Culture performance evaluated in shaker bottles, spinner flasks and bioreactors was comparable, withtypical growth rates in the order of 1.5-fold expansion per day, confirming straightforward widespread scale-up and applicability. Metabolic activity as assessed by the moles lactate produced to glucose consumed was 1.7, consistent with a primarily glycolytic metabolism. Adaptation times for hPSCs transitioning from 2D to 3D aggregate culture varied with different hPSC lines with typically one passage in 3D suspension culture conditions required before a consistent and high expansion passage over passage was obtained. Additionally, protocols were developed for use on a Hamilton® robotic platform for reproducible, matrix-free, high-throughput hPSC suspension culture at a small scale (96-well format). mTeSR™3D enables efficient scale-up and scale-down of hPSC cultures with a greatly simplified workflow.

Optogenetic Investigation of Effects of Beta-Catenin Signaling Dynamics on Neural Stem Cell Differentiation

5th International Conference on Stem Cell Engineering

- Conference: International Conference on Stem Cell Engineering
- Year: 2016
- Proceeding: International Conference on Stem Cell Engineering 2016
- Group: Poster Submissions
- Session:

  Poster session

- Time:

  Tuesday, October 25, 2016 - 5:30pm-7:30pm

Authors:
Rosenbloom, A., University of California, Berkeley
Bugaj, L., UC Berkeley
Kane, R., Rensselaer Polytechnic Institute

Adult neural stem cells (NSC) continuously generate new neurons in specific regions of the brain throughout adulthood to modulate learning and memory, and represent therapeutic targets due to their potential roles in the pathology of diseases and promise in cell replacement therapies. In general, stem cell niches are likely highly dynamic, and the NSC niche in particular likely presents signaling molecules that vary in concentration and duration at timescales ranging from electrophysiological activity, to circadian rhythm, to organismal behavior and environment. Wnt signaling activates Î²-catenin to induce neuronal differentiation of adult hippocampal neural stem cells *in vivo* and *in vitro*. To address the question of how dynamics in signaling impacts cell function, our lab developed a tunable optogenetic system to modulate
Î²-catenin signaling via Cry2 oligomerization of the LRP6 intracellular domain. Similar to responses observed upon Wnt3a ligand addition, blue light illumination of NSCs expressing Cry2-LRP6c induces neuronal differentiation. This outcome enables one to pose many questions related to signaling dynamics, including: do stem cells differentiate when the integral of signaling during a given temporal window exceeds a key threshold,

or do dynamics in signal presentation matter? Continuous illumination at different light intensities *in vitro* resulted in a saturable dose response in which neuronal differentiation progressively increased from 5% to 60%. However, variation in signaling intensity over time yielded different results. Specifically, we observed that light initial pulses of variable duration, or oscilllating illumination at frequencies >12 hours, yielded considerably less neuronal differentiation than in cells that received the same overall signal dosage but with continuous illumination. Furthermore, this signal stimulation followed by signal loss led to increased apoptosis, indicating that exposure to this signal rendered cells dependent upon it not only for differentiation but also for survival, potentially offering a general mechanism for removal of incompletely or poorly differentiated stem cells from tissue. Whole transcriptome sequencing (RNA-seq) revealed a several candidate genes that may play significant roles in differentiation and survival, and overexpression of one cell cycle regulatory factor, whose endogenous expression is lost upon Wnt signal withdrawal, significantly decreased apoptosis and rescued
differentiation upon signal loss. These results indicate potentially critical roles for both cell cycle exit and prevention of aberrant cell cycle re-entry upon loss of Î²-catenin activity. In sum, we harness optogenetics to demonstrate that not only the overall dosage of a signal but temporal dynamics in its presentation can exert a strong impact on stem cell behavior, work that offers further insights into the biology and potential translational potential of adult neurogenesis.

*** Formatting Issue - Best Available Paper/Graphic ***

Production and Controlled Release of a Novel Therapeutic Protein for the Treatment of Muscular Dystrophies

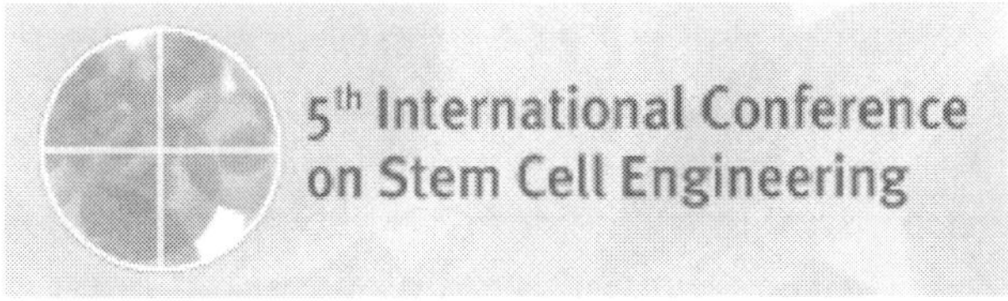

- Conference: International Conference on Stem Cell Engineering
- Year: 2016
- Proceeding: International Conference on Stem Cell Engineering 2016
- Group: Poster Submissions
- Session:

  Poster session

- Time:

  Tuesday, October 25, 2016 - 5:30pm-7:30pm

Authors:
Lev, R., Technion
Kossover, O., Technion
Lati, Y., Technion

Production and controlled release of a novel therapeutic protein for the treatment of muscular dystrophies

Rachel Lev, Olga Kossover, Yonatan Lati, Dror Seliktar

Muscular dystrophies (MDs) are genetic disorders characterized by progressive muscle wasting, leading to limitations in motor capacity and in many cases to progressive paralysis and death1. Cripto, an extracellular protein, has been recently found by our collaborator, Prof. Minchiotti (IGB Naples, Italy), to have therapeutic value in alleviating muscle injury and diseases by acting as a local recruitment factor for stem cells. The local recruitment of progenitor cells facilities muscle regeneration and promotes satellite cell progression toward the myogenic lineage2. A wide range of bio-therapeutic recombinant proteins are currently synthesized using mammalian cell culture methods. However, since protein-based therapeutics may require large doses over a long period of time, manufacturing capacity becomes a critical issue as the drug substance must be produced effectively in large scale for both commercial and clinical requirements. In this study, we improve upon the production capacity of exogenous therapeutic Cripto by using suspension bioreactors in combination with a microgel 3D cultivation system. Subsequently, we have been developing a controlled release delivery system for the recombinantly expressed Cripto protein. Semi-synthetic hydrogels made from PEG-fibrinogen (PF) provide the substrate for protein production as well as the controlled release in situ delivery vehicle. For efficient production of Cripto, PF microcarriers were designed to enable mammalian cell survival, proliferation, and secretion of large amounts of therapeutic proteins. Cripto overexpressing HEK 293T cell lines were encapsulated in the PF microcarriers and cultivated in stirred suspension bioreactors. PF microcarriers were prepared to exhibit high mechanical strength in order to resist shear forces and biodegradation associated with longâ term suspension culture in the bioreactors. In addition, a PF micro-carrier delivery system was developed for intramuscular injection of the purified Cripto protein to provide stability and protection during and after the in vivo procedure. Preliminary studies of the *in vitro* controlled release of Cripto from the micro-scale delivery system shows a biphasic release profile, characterized by an initial burst release followed by a sustained release phase. In vivo experiments will soon be performed to examine and characterize the essential properties of our therapeutic protein release system.

In summary, PEG-Fibrinogen microcarriers hold great promise for both large scale protein production and for controlled release of therapeutic proteins.

1. Mercuri E, Muntoni F., *Lancet* 2013; 381:845-860.
2. Guardiola O et al., *Proc Natl Acad Sci USA* 2012; 109:E3231-E32

Putative Mesothelioma Stem Cells May be the Key Target to Inhibit Cancer Cell Repopulation during the Intervals of Chemoradiation in Murine Models

5th International Conference on Stem Cell Engineering

- Conference: International Conference on Stem Cell Engineering
- Year: 2016
- Proceeding: International Conference on Stem Cell Engineering 2016
- Group: Poster Submissions
- Session:

  Poster session

- Time:

  Tuesday, October 25, 2016 - 5:30pm-7:30pm

Authors:
Yun, Z.
Wu, L.
Blum, W.
Pecze, L.
Kohno, M.
Zhao, Y.
Felley-Bosco, E.
Schwaller, B.
de Perrot, M.

Cancer cell repopulation between cycles of chemotherapy is a major factor limiting the efficacy of treatment. Cancer stem cells (CSC) may play critical roles in this process. Our hypothesis is to identify mesothelioma stem cells (MSC) and then make it feasible to design specific target therapy against MSC and improve the efficacy of malignant mesothelioma.

Methods: Murine mesothelioma AB12 and RN5 cells with or without chemoradiation were used to compare tumor incidence in vivo and evaluate gene expression in vitro by real time PCR. Then we developed RN5-EOS-Puro-R cell line by transducing the EOS (Early transposon promoter and Oct-4 (Pou5f1) and Sox2 enhancers) into RN5 cells with lentiviral vector. RN5-EOS knock-in cells with GFP reporter can be selected by puromycin so as to achieve a highly enriched population of putative MSC. Characterization of MSC and RN5-EOS parental cells were investigated. Gene profile was evaluated by microarray assay to screen potential MSC markers.

Results: Tumor growth delay of murine mesothelioma AB12 was achieved after each cycle of cisplatin, however, tumors grew back up rapidly due to cancer cell repopulation between courses of chemotherapy. Strikingly, 10 times less number of irradiated cells in both cell lines can lead to similar tumor incidence and growth rate than untreated cells. The expression of cancer stem cell-associated genes such as CD24, CD133, CD90 and uPAR were up-regulated dramatically after chemoradiation. Highly enriched MSC after selection with puromycin displayed increasing GFP population and had typical property of stemness. Comparatively, the proportion of MSC significantly increased after RN5-EOS parental cells were treated with either chemotherapy, or Î³-ray radiation, or in combination, while MSC showed more resistant to the above treatments. Gene profile identified a group of genes that are most likely MSC specific.

Conclusion: Putative MSC possess property of stemness showing more resistance to chemoradiation, suggesting that MSC may play critical roles in cancer cell repopulation. Further identification of selected genes may be used to design novel target therapy against MSC in murine mesothelioma models. Targeting MSC would be able to eliminate cancer cell repopulation and may be suitable to other types of cancer as well.

Keywords: Mesothelioma stem cell (MSC); Enriched RN5-EOS-Puro-R cells; Cancer cell repopulation; Microarray; Chemoradiation

Superresolution Fluorescence Microscopy Reveals the Nanoscale Architecture of Actin Cytoskeleton and Integrin-Based Adhesions in Mouse Embryonic Stem Cells

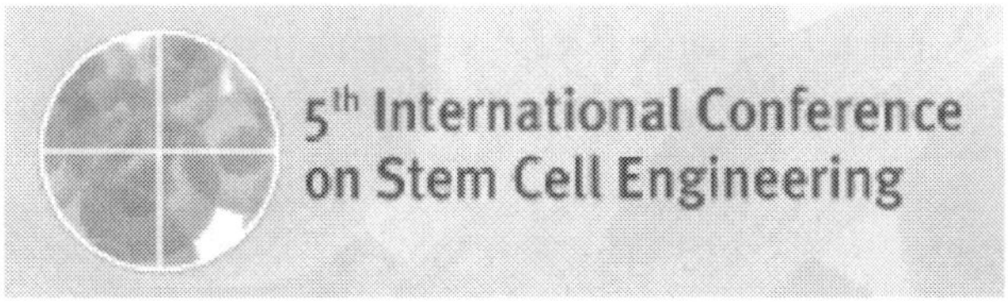

- Conference: International Conference on Stem Cell Engineering
- Year: 2016
- Proceeding: International Conference on Stem Cell Engineering 2016
- Group: Poster Submissions
- Session:

  Poster session

- Time:

  Tuesday, October 25, 2016 - 5:30pm-7:30pm

Authors:
Xia, S., Mechanobiology Institute
Yim, E. K., National University of Singapore
Kanchanawong, P., Mechanobiology Institute

Efficient differentiation of embryonic stem cells (ESCs) into specific cell types is a major focus of regenerative medicine. In addition to soluble biochemical cues, the specific lineage commitment of ESCs also depends on mechanical cues, such as substrate rigidity. However, the underlying mechanisms that couple stem cell mechano-environment sensing to differentiation and maintenance remain obscure. While in many specialized cells, integrin-based adhesion complexes are organized into multi-layer nanostructures that link the extracellular matrix to the actin cytoskeleton and constitute major sites of mechanotransduction, few studies have investigated the structural basis of integrin adhesions and actin cytoskeleton in ESCs. Here we characterize the actin cortex and integrin adhesions in mouse ESC (mESCs) using superresolution fluorescence microscopy. We found that in mESCs, sparse and isotropic actin network comprises the cell cortex, with filopodia and actin bundles confined to the cell periphery. Actin filaments in the cortical meshwork appears to converge into several nodal structures which contain cortactin and a-actinin, while, surprisingly, actin polymerization activities appear to be evenly distributed throughout the cortex. We further characterized the integrin adhesions of ESCs, observing myosin II contractility-dependent localization of hallmark focal adhesions proteins such as focal adhesion kinase, paxillin, talin, vinculin, zyxin, and a-actinin to adhesions formed on fibronectin, laminin, and gelatin. Furthermore, at the nanoscale level these proteins are stratified into a multi-layer architecture, consisting of the integrin-signalling layer, force transduction layer, and actin regulatory layer. Thus, despite the small size, low density, and sparse peripheral distributions, the mESC integrin adhesions are mechanosensitive, compositionally mature, and organized into a conserved nanoscale architecture. Our results suggest that mESCs may adopt an alternate cellular organization paradigm whereby a sparse isotropic cortex is anchored by mechanosensitive integrin adhesions. Their integrated functions may account for the high pliability of ESCs that supports rapid sensing and adaptation to their microenvironment.

The Effect of Maa Polymer-Stimulated Macrophages on Endothelial Cells

5th International Conference on Stem Cell Engineering

- Conference: International Conference on Stem Cell Engineering
- Year: 2016
- Proceeding: International Conference on Stem Cell Engineering 2016
- Group: Poster Submissions
- Session:

  Poster session

- Time:

  Tuesday, October 25, 2016 - 5:30pm-7:30pm

Authors:
Talior-Volodarsky, I., University of Toronto
Zhang, D., University of Toronto
Mahou, R., University of Toronto
Sefton, M. V., University of Toronto

The effect of MAA polymer-stimulated macrophages on endothelial cells.

Ilana Talior-Volodarky[1], DavidZhang[1], Redouan Mahou[1] and Michael Sefton[1]

[1]University of Toronto, Institute of Biomaterials and Biomedical Engineering, Ontario, Canada.

Polymers based on methacrylic acid (MAA) were previously shown to have a therapeutic effect on wound closure through the promotion of angiogenesis of surrounding tissues. However, it is unclear how this polymer exerts these beneficial effects.

The main goal of this current study was to understand and characterize the response of endothelial cells and macrophages to MAA-co-IDA copolymer film (MAA, IDA=isodecyl acrylate, in comparison to methyl methacrylate copolymer MM-co-IDA, used as a control) by identifying a critical signalling pathway, which is activated during cell-material interaction.

MAA upregulated IGF-1 (Insulin like Growth Factor 1), but not IGF-2 (Insulin like Growth Factor 2) in dTHP1cells and mouse macrophages. It was hypothesized that IGF-1 in MAA-conditioned medium stimulated HUVEC migration.

Scratch wound assay in HUVEC indicated increased migration of the cells after their incubation with MAA-treated dTHP1 conditioned medium in comparison to MM-treated dTHP1 conditioned medium. The enhanced effect was inhibited by IGF-1R inhibitor (AG1024).

Gene expression analysis in dTHP1 and M0, M1 and M2 macrophages (BMDM Bone Marrow-Derived Macrophages) showed that MAA polymer increased the expression of IGFBP-3 (3.5-fold in dTHP1; 3.8- fold in BMDM) relative to MM. IGFBP-3 is known to block IGF-1 access to the IGF-1 receptor and thereafter deliver more IGF1 into the blood stream. Macrophages also demonstrated decreased expression of IGF-R1 when treated with MAA polymer. In contrast, HUVEC treated with MAA stimulated dTHP1 conditioned medium showed increased expression of IGF-R1 (1.7-fold), but decreased expression of IGFBP-3, which is opposite to dTHP1 cells treated with MAA polymer film.

Arg1 (M2 marker), was upregulated in both M0 and M1 macrophages treated with MAA polymer compared with control treatment, whereas, the expression of M1 marker - iNOS in M0 and M2 macrophages was unaffected in MAA stimulated conditions. Exposure of MAA treated macrophages to AG1024 inhibitor induced a reduction of Arg1 expression.

We related the *in vitro* findings to the angiogenic response *in vivo*. All *in vivo* results were consistent with *in vitro* studies: upregulation of IGF-1 transcript level in MAA-treated animals was observed; Arginase-1 expression level was significantly increased in animals treated with MAA polymer and iNOS expression was unaffected in MAA stimulated conditions. Most importantly, AG1024 abrogated the increased CD31+ vessels density seen with MAA implants, but had no effect on MM implants.

This data suggest that MAA-co-IDA stimulates an IGF-1 mediated macrophage response to the biomaterial, via macrophages polarization toward the M2 phenotype which in turn activates endothelial cells required for the process of vascularisation.

Understanding the MAA stimulation mechanism can be used to advance the knowledge of this biomaterial and cell or tissue interaction, thereby facilitating the development of MAA polymer for new therapeutic applications, such as revascularization and tissue regeneration.

Key words: Angiogenesis, IGF-1, macrophages and endothelial cells.

Viable Harvest and Washing of Stem Cells Using Unifuge, Contnous Single Use Centrifuge

5th International Conference
on Stem Cell Engineering

- Conference: International Conference on Stem Cell Engineering
- Year: 2016
- Proceeding: International Conference on Stem Cell Engineering 2016
- Group: Poster Submissions
- Session:

  Poster session

- Time:

  Tuesday, October 25, 2016 - 5:30pm-7:30pm

Authors:
Richardson, D., PSA
Summary: 10 liters of myeloid progenitor suspension cells were grown in a Wave bioreactor. The cells were aseptically collected with the UniFuge single use centrifuge. The viability of the grown cells measured 95%. The suspension cells were grown for 4 days, doubling each day. The diameter size of the cells are 15 -20 micron. The cells reached a final cell density 4.93 million cells/ml at time of harvest. To harvest the cells, the UniFuge centrifuge parameters were feed rate (LPM) and G force . The cells were layered on a PBS buffer; a manual prefill step was utilized by switching supply lines. The suspension cells were collected aseptically. After Harvest, the cells were frozen for 24 hours, thawed and viabilities measured.

Conclusion: For the Viable collection myeloid progenitor like suspension cells, the UniFuge Single use centrifuge **met all three objectives**. 1. Collect 93% of cells at 700 x G. 2. Cells maintained > 90 viability after 24 hour thaw. 3. Prefill bowl with PBS buffer prior to feed.

Viable Stem Cell Harvest and Wash with Continous Sinlge Use Centrifuge, Unifuge

5th International Conference on Stem Cell Engineering

- Conference: International Conference on Stem Cell Engineering
- Year: 2016
- Proceeding: International Conference on Stem Cell Engineering 2016
- Group: Poster Submissions
- Session:

  Poster session

- Time:

  Tuesday, October 25, 2016 - 5:30pm-7:30pm

Authors:
Richardson, D., PSA
Summary: 10 liters of myeloid progenitor suspension cells were grown in a Wave bioreactor. The cells were aseptically collected with the UniFuge single use centrifuge. The viability of the grown cells measured 95%. The suspension cells were grown for 4 days, doubling each day. The diameter size of the cells are 15 -20 micron. The cells reached a final cell density 4.93 million cells/ml at time of harvest. To harvest the cells, the UniFuge centrifuge parameters were feed rate (LPM) and G force . The cells were layered on a PBS buffer; a manual prefill step was utilized by switching supply lines. The suspension cells were collected aseptically. After Harvest, the cells were frozen for 24 hours, thawed and viabilities measured.

Conclusion: For the Viable collection myeloid progenitor like suspension cells, the UniFuge Single use centrifuge **met all three objectives**. 1. Collect 93% of cells at 700 x G. 2. Cells maintained > 90 viability after 24 hour thaw. 3. Prefill bowl with PBS buffer prior to feed.

AUTHOR INDEX

AUTHOR INDEX

AUTHOR INDEX